SYMBOLIC MODEL CHECKING

SYMBOLIC MODEL CHECKING

by

Kenneth L. McMillan
Carnegie Mellon University

Kluwer Academic Publishers
Boston/Dordrecht/London

Distributors for North America:
Kluwer Academic Publishers
101 Philip Drive
Assinippi Park
Norwell, Massachusetts 02061 USA

Distributors for all other countries:
Kluwer Academic Publishers Group
Distribution Centre
Post Office Box 322
3300 AH Dordrecht, THE NETHERLANDS

Library of Congress Cataloging-in-Publication Data

McMillan, Kenneth L.
  Symbolic model checking / by Kenneth L. McMillan.
      p. cm.
  Includes bibliographical references and index.
  ISBN 0-7923-9380-5
    1. Electronic digital computers--Circuits--Design--Data
  processing. 2. Symbolic circuit analysis--Data processing.
  3. Logic design--Data processing.   I. Title.
  TK7888.4.M43 1993
  621.39'2--dc20   93-24859
                                                             CIP

Copyright © 1993 by Kluwer Academic Publishers

All rights reserved. No part of this publication may be reproduced, stored in a retrieval system or transmitted in any form or by any means, mechanical, photo-copying, recording, or otherwise, without the prior written permission of the publisher, Kluwer Academic Publishers, 101 Philip Drive, Assinippi Park, Norwell, Massachusetts 02061.

Printed on acid-free paper.

Printed in the United States of America

Dedicated to the memory of William L. McMillan

CONTENTS

FOREWORD ix

PREFACE xiii

1 **INTRODUCTION** 1
   1.1 Background 3
   1.2 Scope of this work 9

2 **MODEL CHECKING** 11
   2.1 Temporal logic 13
   2.2 The temporal logic CTL 16
   2.3 Fixed points 18
   2.4 CTL model checking 20

3 **SYMBOLIC MODEL CHECKING** 25
   3.1 Boolean representations 25
   3.2 Symbolic models 26
   3.3 Binary Decision Diagrams 31
   3.4 Examples 39
   3.5 Graph width and OBDDs 49

4 **THE SMV SYSTEM** 61
   4.1 An informal introduction 63
   4.2 The input language 69
   4.3 Formal semantics 78

5 **A DISTRIBUTED CACHE PROTOCOL** 87

5.1	The Protocol	89
5.2	Verifying the protocol	100
5.3	Discussion	111

6 MU-CALCULUS MODEL CHECKING 113
6.1	The Mu-Calculus	113
6.2	Symbolic models	115
6.3	Symbolic algorithm	116
6.4	Applications of the Mu-Calculus	117
6.5	Related research	122

7 INDUCTION AND MODEL CHECKING 129
7.1	The general framework	130
7.2	Induction and symbolic model checking	132
7.3	Example: The Gigamax protocol	134
7.4	Induction in other models	139
7.5	Related research	140

8 EQUIVALENCE COMPUTATIONS 143
8.1	State equivalence	143
8.2	Methods for functional composition	146
8.3	Experimental results	148

9 A PARTIAL ORDER APPROACH 153
9.1	Unfolding	155
9.2	Truncated unfoldings	163
9.3	Application example	168
9.4	Deadlock and occurrence nets	171
9.5	Conclusion	174

10 CONCLUSION 179

REFERENCES 183

INDEX 191

FOREWORD

Logical errors in sequential circuit designs and protocols are an important problem for hardware designers. Such errors can delay getting a new product on the market or cause the failure of a computer system that is already in use. The most widely used method for verifying such systems is based on extensive simulation and can easily miss significant errors when the number of possible states of the system is very large. Although there has been considerable research on the use of theorem provers, term-rewriting systems, and proof checkers to verify such systems, these techniques are time consuming and often require a great deal of manual intervention.

Over the past ten years, my research group at Carnegie Mellon University has developed an alternative approach to verification called *temporal logic model checking*. In this approach specifications are expressed in a propositional temporal logic, and circuit designs and protocols are modeled as state–transition systems. An efficient search procedure is used to determine automatically if the specifications are satisfied by the transition systems. The initial idea for model checking appeared in a paper that I wrote with my first graduate student, Allen Emerson, in 1981. The earliest example that we verified using this technique was a version of the *alternating bit protocol* with 251 states. Neither of us could have possibly predicted the size of the transition systems that are now routinely checked by this method.

The main disadvantage of the original model checking algorithm was the *state explosion problem* which occurred if the system being verified had many components that could make transitions in parallel. Because of this problem, many researchers in formal verification predicted that model checking would never be practical for large circuits and protocols. This view was reinforced by early implementations which could only handle transition systems with a few thousand states and, therefore, were only useful for verifying systems with a small number of components. As late as 1987, no one could verify transition systems with more than a million states by using model checking techniques.

During the last five years, the size of the transition systems that can be verified by model checking techniques has increased dramatically. The initial breakthrough was made in the fall of 1987 by one of my graduate students, Ken McMillan, who realized that the explicit adjacency-list representation that we had been using for transition systems severely limited the size of the circuits and protocols that we could verify. By representing transition systems implicitly using Binary Decision Diagrams (BDDs), he was able to handle some examples that had more than 10^{20} states. McMillan made this observation independently of the work by Coudert and Madre on using BDDs for checking equivalence of deterministic finite state machines. In fact, he made the key observation during the first six weeks of his first year as a graduate student at CMU. Refinements of the BDD-based techniques by two other graduate students, Jerry Burch and David Long, have pushed the state count up to more than 10^{100}. By combining model checking with various abstraction techniques, my research group has been able to handle even larger circuits and protocols. In one example, we were able to verify a pipelined ALU with more than 10^{1300} states.

For this insight McMillan was a cowinner of the 1992 ACM Doctoral Dissertation Award. This book is his Ph.D. thesis. In my opinion, it is the best introduction to temporal logic model checking that has been written. The SMV model checker described in this thesis is based on a dataflow-oriented hardware description language which can be annotated by specifications expressed in the temporal logic CTL. The model checker extracts a transition system from a program in the SMV language and uses a BDD-based search algorithm to determine whether the system satisfies the CTL specifications. If the transition system does not satisfy some specification, the verifier will produce an execution trace that shows why the specification is false. These traces are particularly useful in finding and correcting errors in complex systems.

The SMV system has been distributed widely, and many circuits and protocols have now been verified with it. Perhaps, the most impressive example of its use is the verification of the cache coherence protocol described in the IEEE Futurebus+ standard (IEEE Standard 896.1–1991). Although development of the Futurebus+ cache coherence protocol began in 1988, all previous attempts to validate the protocol were based entirely on informal techniques. In the summer of 1992 David Long constructed a precise model of the protocol in the SMV language and then used McMillan's model checker to show that the resulting transition system satisfied a formal specification of cache coherence. He was able to find a number of previously undetected errors and potential errors in the design of the protocol. To the best of my knowledge this is the first time that an automatic verification tool has been used to find errors in an IEEE standard.

Foreword

I believe that the examples in this thesis provide convincing evidence that SMV can be used to debug real industrial designs. Nevertheless, much work remains to be done to realize the full potential of this new verification technique. In addition to describing the most powerful model checking system that has been developed, McMillan's thesis provides an excellent introduction to these new areas of research. Certainly, the most important problem is to characterize the class of circuits that have small BDD representations and, therefore, can be verified by symbolic model checking techniques. McMillan gives what I think is the best characterization to date of such circuits. From the examples that we have considered already, it is clear that use of abstraction and induction will be crucial in reasoning about large systems in the future. McMillan's thesis illustrates how these techniques can be applied in a number of realistic examples. His thesis also demonstrates convincingly that the state explosion problem is reduced if the possible executions of a concurrent system are modeled by partially ordered event structures instead of by execution traces or computation trees with interleaving of events. Several other important research topics are discussed in addition to the three that I have listed. Because his thesis opens up so many exciting areas for future research, I am positive that it will continue to have a profound influence in research on formal verification methods in the years to come.

Edmund M. Clarke, Jr.

Carnegie Mellon University

PREFACE

This book is a revised edition of my doctoral thesis, submitted in 1992 to Carnegie Mellon Univerity. As such, it is chiefly a report of research in the area of automatic formal verification. However, I hope that it can also serve as an introduction to the area for those interested in applying the symbolic model checking technique to verify their own designs, or in evaluating the method with a view to deciding what verification tools might be integrated into their own design systems. Thus, I have not assumed familiarity with such topics as temporal logic or model checking (though certainly a more thorough treatment of these topics can be obtained from other sources). In addition, I have tried to keep the emphasis on the practical side, using realistic examples whenever possible. The most involved of these is the verification of the cache consistency protocols of the Encore Gigamax multiprocessor, to which a chapter is devoted. Those readers interesting in applying the methods will find chapters 1-5 to be of the greatest interest. On the other hand, chapters 6-9 are more oriented toward researchers in the field. These chapters cover research topics that are more open-ended, and less clear in their implications. They will also require, I think, a greater degree of patience and mathematical fortitude on the part of the reader in order to decipher them. Nonetheless, chapter 7 in particular, on induction, may be of interest to the practically minded, since it extends the treatment of the cache protocol example to cover systems of arbitrary size. I hope that the presentations in chapters 6-9 may stimulate others to extend the work.

The contents of the book can be summarized roughly as follows: The concept of formal verification and the state explosion problem are introduced in chapter 1. Very briefly, if a system can be modeled as a finite state machine, we can very efficiently prove properties about that system expressed in temporal logic by a technique called model checking, the subject of chapter 2. Unfortunately, finite state models of *concurrent* systems (*eg.*, computer hardware) grow exponentially in size as the number of components of the system increases. This is known widely as the *state explosion problem*, and has limited finite state verification methods to small systems.

Chapter 3 introduces a method of skirting this problem called *symbolic model checking*. This technique avoids constructing the finite state model by representing it symbolically as a Boolean formula. This allows efficient methods of manipulating Boolean algebra to be applied to the model checking process. In particular, a representation called Binary Decision Diagrams can be applied. A significant result from chapter 3 identifies a structural class of sequential circuits which can be represented for the purpose of model checking with Binary Decision Diagrams of low complexity. This is born out by experimental results on example circuits and systems, including the Encore cache protocol.

In chapter 4, a language is developed for describing sequential circuits and protocols at various levels of abstraction. This language has a synchronous dataflow semantics, but allows nondeterminism and supports interleaving processes with shared variables. A system called SMV can automatically verify programs in this language with respect to temporal logic formulas, using the symbolic model checking technique.

Chapter 5 describes the verification of the cache protocol, expressing it in the SMV language, detailing what properites can be proved, and explaining the performance of the model checking algorithm in terms of the theory developed in chapter 3. The symbolic model checking technique revealed subtle errors in this protocol, resulting from complex execution sequences that would occur with very low probability in random simulation runs. This highlights an important aspect of model checking – the ability to generate counterexamples.

Chapter 6 generalizes the symbolic model checking technique from the simple temoral logic described in chapter 2, to a much more expressive logic called the Mu-Calculus. Several application of the Mu-Calculus are described.

Chapter 7 develops a technique that uses model checking to prove properties of systems of arbitrary size (generated by simple rules). The proof is checked automatically, but the user must supply a special model called a *process invariant* to act as an inductive hypothesis. A process invariant is developed for the distributed cache protocol, allowing properties of systems with an arbitrary number of processors to be proved.

Chapter 8 deals with the question of computing the state equivalence relation between finite state machines, relying heavily on the material in chapter 5. Benchmark results using a variety of optimizations are presented.

Finally, in chapter 9, an alternative method is developed for avoiding the state explosion in the case of asynchronous control circuits. This technique is based

Preface

the unfolding of Petri nets, and is used to check for hazards in a distributed mutual exclusion circuit.

This book came about because of the influence, advice and assistance of a large number of people, whom I would like to thank. By far the most significant influence on the work is that of Ed Clarke, my advisor at CMU. His work is the foundation of almost everything in it. Ed got me started in the field by teaching me what I know about verification, and opening doors for me in the research community. His enthusiasm for my early, somewhat untutored efforts in the field gave me the confidence to pursue my own ideas. Another important influence on this work and its author is Bob Kurshan. In more than once instance, his insistence on solving a particular problem led to a general solution in an unexpected way.

The work described in this book also rests rather heavily on that of Randy Bryant. To a certain extent, the symbolic model checking technique resulted from the fortunate coincidence of my being at the same institution as Randy. Finding an application for this technique was also a more or less serendipitous occurrence, and for this I have to thank the people at Encore Computer Corporation, including Dyung Van Le, Jim Schwalbe and Drew Wilson, for taking an interest in formal verification, and generously allowing me to use their system as an application (and to publish it no less!).

I must also thank David Long and Jerry Burch, who contributed substantially to the ideas in this book. I owe special thanks to David, who is a font of information, and who helped me to solidify ideas in countless discussions, when no doubt he had more important things to do.

Thanks to Robert Brayton and the CAD group at the University of California at Berkeley, who gave my thesis a very careful and intelligent reading, and to Allan Fisher. It was always a pleasure to have discussions with Allan, and no doubt some of his insights can be found interspersed in these pages. Of course, I am indebted to all the good people who make the Carnegie Mellon School of Computer Science and its facilities work. CMU provides a unique environment for graduate students, and I consider myself privileged to have been a part of it.

Finally, to Tracy Slein, the one indispensible person in the whole process – I can't thank you enough.

SYMBOLIC MODEL CHECKING

1
INTRODUCTION

Formal verification means having a mathematical model of a system, a language for specifying desired properties of the system in a concise, comprehensible and unambiguous way, and a method of proof to verify that the specified properties are satisfied. When the method of proof is carried out substantially by machine, we speak of automatic verification. This book deals with methods of automatic verification as applied to computer hardware.

The practical motivation for study in this area is the high and increasing cost of correcting design errors in VLSI technologies. There is a growing demand for design methodologies that can yield correct designs on the first fabrication run. Moreover, design errors that are discovered before fabrication can also be quite costly, in terms of the engineering effort required to correct the error, and the resulting impact on development schedules. Aside from pure cost considerations, there is also a need on the theoretical side to provide a sound mathematical basis for the design of computer systems, especially in areas that have received little theoretical attention. Every computer engineering student learns the classical methods of switching theory – how to design and optimize combinational logic and finite state machines using Boolean algebra – but there are no mathematical methods available for dealing with the complex *systems*, such as cache consistency and network protocols, that are being built into hardware today.

As a result, the engineer using current design methods is faced with two ill-characterized and increasingly intractable problems. The first is to create a simulation regime with test patterns sufficient to expose any logical errors in the design. The second is to interpret the vast quantity of simulation output to determine whether incorrect behavior has occurred, in the absense of precise and unambiguous specifications. Formal verification addresses both of these issues, by encompassing all possible behaviors of the model, and by providing a formal language for specification. What formal verification cannot do, however, is guarantee an irrefutably and unassailably correct system. Any logical proof is only good insofar as one has confidence in the assumptions on which it is based and the desirability of the proved results. Hence, a specification which is proved but incomprehensible is worthless. The goal of formal verification should be the same as other types of design verification – to avoid costly design revisions and catastrophic failures, and to increase one's understanding of the design issues. Ideally, formal verification should be applicable at the earliest stages of the design process, to prevent prevent errors from being propagated from the conceptual level into the detailed design.

There are two more or less distinct traditions in formal verification of computer hardware. One is based on a long history of work in automatic theorem proving, and the other is based on somewhat newer paradigm called *model checking*. In the former case, one expresses the system model and the specifications in a suitable logic, and constructs a proof in the logic that the system model implies the specifications. This is a very powerful and flexible approach, but unfortunately it can involve generating and proving literally hundreds of lemmas in painstaking detail. Even a very mathematically oriented designer is unlikely to find this more appealing than the admittedly inadequate method of verification by simulation. Model checking, on the other hand, is more limited in scope, but is fast and fully automated. The system model is in essence a finite state machine, and specifications are written in a specialized language called a propositional *temporal logic*. These logics are limited with respect to the very powerful logics handled by general theorem provers, but are quite simple and concise, and can express a wide variety of useful properties (usually categorized as safety, liveness, fairness, *etc.*). Examples abound of small but subtle circuits that have been verified in this way, as well as bugs that have been found in published circuits.

Unfortunely, modeling complex systems as finite state machines has an inherent disadvantage which is commonly known as the *state explosion problem*. This is the exponential relation of the number of states in the model to the number of components of which the state is made. Thus model checking is capable of dealing efficiently with very large finite state machines, but not with systems that are made up of a large number of small finite state machines, nor with systems that manipulate data.

There is, however, a fair amount of middle ground to be explored. Here, for example, we take the model checking approach to verification, but attempt to find

Introduction

ways of avoiding the state explosion problem for systems with certain kinds of structure. The approach may at first seem counterintuitive – we return the system model to the realm of symbolic logic, representing it by a Boolean formula. The intent is not to apply traditional deduction methods, however. Instead, it is to combine very powerful algorithms for manipulating Boolean algebra (based on Binary Decision Diagrams [Bry86]) with the fixed point algorithms of model checking. The result is called *symbolic model checking*. One might argue that this is really introducing new algorithms for theorem proving in a restricted setting – in fact symbolic model checking might well be integrated into a general prover as a tactic. The point is that symbolic model checking makes the compromise between expressiveness and the difficulty of automatic proof in a new way. This makes it possible to treat some systems that would be very difficult to handle with either the pure theorem proving or model checking approaches.

In particular, this work examines in detail the modeling and verification of a distributed cache consistency protocol designed for a commercial multiprocessor at Encore Computer Corporation. In the process, the necessary verification theory and algorithms are developed, as well as a language for modeling systems of this type at a suitable level of abstraction. The cache protocol coincidentally serves as an excellent illustration of how properties of Binary Decision Diagrams determine the kind of system structures that can be efficiently verified using symbolic model checking. Results of this kind can serve as a rule of thumb for deciding when symbolic model checking is an appropriate strategy for verification.

1.1 BACKGROUND

Before proceeding, it will be worthwhile to consider briefly some of the historical background that underlies this work. Hardware verification, first of all, has certain elements in common with the more venerable area of program verification, a subject which has an extensive literature. In this regard, computers are most similar to what Pnueli has characterized as *reactive* programs [Pnu86], in that they receive input and produce output in a continuous interaction with their environment, rather than computing a single result and halting. In addition, the behavior of circuits is concurrent in the extreme, since every gate in the system is simultaneously evaluating its output as a function of its inputs. For reasoning about concurrent and reactive programs, Pnueli proposed the use of a formal system originally studied by philosophers, called temporal logic [Pnu77, Pnu86, MP81, Kro87].

1.1.1 Temporal logic

In a temporal logic, the usual operators of propositional logic are augmented by *tense operators*, which are used to form statements about how conditions change in time. One can assert, for example, that if proposition p holds in the present, then proposition q holds at some instant in the future, or at some instant in the past. The tense operators can be combined to express fairly complex statements about past, present and future. For example "if p holds in the present, then at some instant in the future, p will have held in the past." A temporal system provides a complete set of axioms and inference rules for proving all validities in the logic for a given model of time, such as partially ordered time, linearly ordered time, dense time, and even branching time.

Temporal logic can be used to define a semantics for programs which captures not only the traditional pre- and post-conditions of Floyd-Hoare logic, but also a wide variety of temporal properties of programs, such as termination, possible termination, termination under fair scheduling of concurrent processes, *etc.* [CE81a, BAMP81]. It was for this purpose that temporal logic first genertated interest in the verification community. In the hardware area, Malachi and Owicki used temporal logic to give a concise specification of the conditions necessary for an asynchronous circuit to be *speed-independent* [MO81], meaning that its function is independent of the propagation delays of the gates. Bochmann used temporal logic to give a semantics for self-timed circuits, and used this system to verify a corrected version of an arbiter circuit [Sei80a]. Formal proofs of this kind are extremely tedious and difficult, however, and computationally intractable to automate. To simplify the hand proof, Bochmann used a somewhat oversimplified semantics for the circuit elements (neglecting gate delay) and as a result, missed a bug in the design, which was demonstrated by Dill [DC86]. This is a good illustration of the point made earlier that a proof is only as good as the assumptions on which it is based.

The *model checking* technique was introduced by Clarke and Emerson [CE81b] and independently by Quielle and Sifakis [QS81]. Instead of proving the validity of a logical formula for all models, a model checker determines the truth value of the formula in a specific finite model. For branching time logic, the model checking problem is computationally tractable, even though the validity problem is intractable. Here an important distinction between hardware and software systems comes into play – hardware systems are finite-state. This

Introduction 5

allows the proof procedure to be automated using model checking, while maintaining the formal elegance of temporal logic for specifying correct behavior.

The algorithm of Clarke and Emerson first builds a complete state graph of the system model from a description in an appropriate language. The truth value of a formula in the logic is determined by an algorithm which propagates formulas in this state graph until a fixed point is reached. Besides being fast and fully automatic, this technique has the advantage that it can produce state sequences as counterexamples when the formula being checked is false. This has made it possible to find bugs in a number of small but fairly subtle circuit designs [BCDM86, BCD86], including the one verified by Bochmann.

For linear time temporal logic, a procedure was developed for determining the validity of a formula by translating the formula into an automaton by means of a *tableau* construction [RU71, CE81b, BAMP81]. This construction is related the the semantic tableaux method of constructing proofs in standard logic [Smu68]. Each state in the tableau is associated with a set of formulas which are true in that state. Since the number of states in the tableau is exponential in the size of the formula, the method is not practical for proofs about very large systems. However, the tableau method can be used in a model checking framework, yielding an algorithm which is exponential in the size of the formula but linear in the size of the model [LP85].

1.1.2 Automata theoretic methods

In parallel with the work on temporal logic, a number of verification methodologies were developed which use finite state machines to represent both the system model and the specification. In this case verification means comparing the externally visible behaviors of the two machines, usually in terms of the possible sequences of communications between processes. For example, in the s/r model of Kurshan [Kur86], these communication sequences are defined by the language of an ω-automaton (an automaton on infinite strings). Correctness is framed as the containment of the language of the system automaton in the language of the specification automaton. This asymmetric relation makes it possible to underspecify a system, that is, to leave some choices open to the designer. For example, the specification automaton may accept many sequences corresponding to a given input sequence, while the implementation accepts only one such sequence. The use of automata on infinite strings makes it possible to express *liveness properties*. Liveness properties deal with eventualities – events which must occur at some finite but unbounded time. For instance, one can easily construct an automaton to represent all infinite communication sequences such that every time a message is sent on channel A, it is eventually received on channel B. Language containment between ω-automata can be proved by an algorithm which searches for cycles in the state space of a product automaton.

Van de Snepscheut [vdS83] and Dill [Dil88] have used trace theory to model speed independent circuits [Sei80b]. A trace is simply a history of the communications between a process and its environment. The trace sets of two process can be combined in a way which models communication between the two processes by synchronizing signals sent and received on the same channel. Dill's system is a circuit algebra which has both a structural interpretation (describing the physical connection of wires) and a trace theoretic interpretation (describing the communications along those wires). The actual trace sets are determined by the languages of finite automata (in this case, automata on finite strings, hence liveness cannot be modeled). A relationship called *conformance* between two processes determines when one process can safely be substituted for the other in all environments. Conformance can be tested by a polynomial algorithm which searches the state space of a finite automaton derived from the two processes.

In the Calculus of Communicating Systems (CCS) [Mil80], Milner takes a different approach in which external behavior is modeled by a tree rather than a set of sequences. The way CCS treats communication is not well suited to modeling hardware, however, because in CCS a signal cannot be sent until a receiver is ready to receive it. In hardware, on the other hand, a receiver cannot generally prevent a signal from being sent. Also, in CCS, communication is always between two processes, while in hardware signals are often broadcast to many receivers. A calculus specialized to circuits called CIRCAL [Mil83] was developed to remedy these problems. The notion of correctness in process calculi is called *observational equivalence*, meaning that an observer cannot distinguish between two processes by any experiment. This notion of correctness is extremely strict, since it doesn't allow the specifier to leave any choice to the designer regarding the externally visible behaviors. Observational equivalence can be proved by establishing a relation called *bisimulation* between the two processes. For finite state processes, there is a polynomial time algorithm for bisimulation which is very similar to the coarsest partitioning algorithms used for state machine minimization [NH84].

All of these methods can be viewed as variations on the theory of finite automata, tailored for modeling a particular class of systems. In fact, the automata theoretic approach is not very far from the temporal logic approach

Introduction

in practice. The difference is mostly a question of notations, since the *tableau* method provides a way of translating a temporal logic formula into an automaton. Although temporal logic is not as expressive as automata are in characterizing sequences, it was shown by Wolper that temporal logic can be extended using right linear grammars to make it as expressive as automata without increasing the complexity of the decision procedure [Wol83]. Clarke and Kurshan have also proposed a branching time logic in which the temporal operators are defined by finite ω-automata [CGK89].

The verification approaches mentioned above are closely allied to model checking, and all suffer from the state explosion problem. In theory, a variant of the symbolic model checking technique can be applied to each of them, by expressing the necessary correctness condition in a fixed point logic called the Mu-Calculus and using a symbolic algorithm for model checking in this logic.

1.1.3 Reduction

Since the state explosion problem is ubiquitous in finite state verification methods, it is not suprising that a many researchers have studied it. The most common approach to the problem is based on *reduction* – transforming the verification problem to a similar problem in a smaller state space. This is generally done by replacing processes in the model by smaller processes that have similar or identical communication behavior. The most general framework for this kind of reduction is that of Kurshan [Kur87]. Using *homomorphic reductions* of ω-automaton models, it is possible to simplify not only the internal state of a process, but also its external communications. In this methodology, one generally builds a hierarchy of reductions, in which processes at the lowest level are reduced, then combined at the next higher level and further reduced, *etc.* Kurshan advocates building this hierarchy from the top down, so that the most abstract models can be verified before details are filled in at the next lower level.

A hierarchical approach was also taken by Dill in his trace theoretic system for speed independent circuits [Dil88]. In this case, the reduction is obtained mostly by hiding internal signals of a process. There is no provision for abstracting the signals by which the module communicates with its environment. That is, communication always remains at the same level, that of digital signal transitions.

The reduction approach is generally difficult to automate. Usually, the reduced process is obtained in an *ad hoc* manner, and the validity of the reduction is then tested automatically. Some methods have been proposed for obtaining reduced processes automatically, however. For example, in a method called *compositional model checking*, a state minimization procedure is used to obtain a reduced process that is equivalent to the original process with respect to observation via its inputs and outputs [CLM89b, CLM89a]. This reduction preserves the truth value of all formulas in a suitable logic. Graf and Steffen have also studied minimization with respect to a suitable notion of equivalence as a reduction technique [GS91]. Minimization techniques are fairly strict in terms of the required relation between the original and reduced processes, however. As a result, the reduction that can be obtained using these techniques is not generally as great can be obtained using more flexible but unautomated methods.

The symbolic model checking technique is not really an alternative to reduction methods, but is complementary to them. In general, the larger the state space that can be searched automatically, the less the need for reduction. For example, Dill used a reduction (constructed by hand) to verify a speed independent distributed mutual exclusion ring circuit [Dil88]. Using symbolic model checking, there is no need for a reduction – the verification time is polynomial in the size of the ring (cf. chapter 2). On the other hand, symbolic model checking techniques can be used to implement the validity test for reductions (cf. chapter 5), hence the two techniques can be combined.

1.1.4 Induction

In systems of many identical processes, it is sometimes possible to reduce an arbitrary number of processes to a single process while preserving certain properties of interest. For example, Browne, Clarke and Grumberg proposed a reduction technique of this sort which preserves the truth value of formulas in a restricted logic with process quantifiers [BCG86]. A process quantifier allows us to state, for example, that there exists some process in the system which is in a given state. Unfortunately, the reduction, a form of bisimulation, had to be established by hand. There was no automated way of checking it. Kurshan and McMillan proposed an inductive method of establishing the reduction that could be checked automatically [KM89]. Their method is less restrictive in terms of the properties that can be proved since it does not require equivalence. A similar method was described independently by Wolper and 'Lovinfosse [WL89]. Another inductive technique has been described by Shtadler and Grumberg [SG89]. This technique is somewhat more flexible in

Introduction 9

that it treats networks generated by context free grammars, but is limited to bisimulation as a reduction technique. Not surprisingly, induction methods can be combined with symbolic model checking. Chapter 7 develops such a technique and describes its application to verifying properties of the Encore cache consistency protocol that are independent of the number of processors in the system.

1.2 SCOPE OF THIS WORK

The foregoing should make it clear that symbolic model checking is not a new paradigm for verification. Rather, it is a computational technique that can be applied in a wide range of existing verification frameworks, for a variety of purposes. It is prudent, however, to develop the theory in terms of the simplest of these applications – model checking for branching time temporal logic. This is the subject of the next two chapters. Since the method is an attempt to balance computational techniques and logical expressiveness, we will need concrete examples to demonstrate its area of applicability. From these, we will turn to a more theoretical approach, characterizing a class of systems that can be represented efficiently using Binary Decision Diagrams, and hence may be appropriate applications for symbolic model checking.

In the following chapters, we proceed to develop a language for describing system models, with a precise semantics relating programs to their expression as Boolean formulas. This language is used to describe the Encore cache consistency protocol, which in turn is used to illustrate the advantages of symbolic model checking, and how our theoretical results about Binary Decision Diagrams apply to real systems. Following this, we develop the theory necessary to apply symbolic model checking to the various verification frameworks described above. This subject is unified by expressing all of the appropriate verification conditions in a logic called the Mu-Calculus, using a general symbolic algorithm for Mu-Calculus model checking. This provides a suitable framework for two ensuing chapters, which describe the use of symbolic model checking in inductive proofs, and some experiments in computing equivalence relations between finite state machines. A final chapter, not directly related to symbolic model checking, is included in the hope that it will contribute to an understanding of the nature of the state explosion problem. It deals with the question of how independence of events in concurrent systems contributes to the state explosion problem, and how a partially ordered representation can avoid this problem.

2

MODEL CHECKING

As mentioned in the introduction, a formal verification system has several basic elements. First, we require a *model*. A model is an imaginary universe, or more generally, a class of possible imaginary universes. To make our model meaningful, we require a *theory* that predicts some or all of the possible observations that might be made of the model. An observation generally takes the form of the truth or falsehood of a predicate, or statement about the model. Finally, to verify something meaningful about the model, we require a *methodology* for proving statements that are true in the theory.

In program proving, the universe is a totally imaginary one, driven by mechanisms (the compiler and hardware) of which the programmer need have no knowledge. A logician is free to assign any semantics at all to programs, provided a compiler writer and hardware designer agree to implement them. This makes program proving an artificial science, in the sense that the theory we assign it is valid by construction. In contrast, a hardware verification system requires a model of a real physical system. The underlying physical mechanism is still invisible to us (we can only postulate its existence), but we can empirically construct a model which predicts the necessary observations with a sufficient degree of accuracy for our purposes (the verification of digital circuits). It turns out that the required degree of accuracy is not very large. Though quite accurate models are possible (for example, using partial differential equations to describe the time evolution of fields and particle densities) a suitable design style makes it possible to consider only the digital (one or zero) value of voltages, ignoring entirely the exact voltage within the digital ranges, and the time it takes to switch from one range to another. Depending on the design style (*eg.*, synchronous or self-timed), different models may be appropriate. In certain rare cases, we may have to use differential equations to model the analog behavior of circuits (for example, when metastability arises). In this work, though, we will consider only fairly abstract models of circuits as finite state machines. Thus, we return to the science of the artificial, wherein we choose the theory to suit our needs, but with the understanding that a method exists for translating our models into real systems.

The kind of theory that emerges for the model depends to a large extent on the kind of experiments the observer is able to perform. For example, in traditional program proving systems, the observer is allowed to set up the initial state of the program, wait for the program to terminate, and then examine the final state. The theory of this model can be expressed in a kind of before-and-after logic whose axioms determine the semantics of programs. For example, in Floyd-Hoare logic [Hoa69], the formula

$$\{\text{true}\}\ x := y\ \{x = y\}$$

is an axiom: for any initial condition, after the program $x := y$ terminates, x and y have the same value. The fact that no other variables change value in the process can also be expressed as an axiom:

$$\{z = a\}\ x := y\ \{z = a\}$$

provided neither z nor a depend on x.

In this system, if the program fails to terminate (diverges), the observer must simply wait forever, $ie.$, no observation is possible. One might ask whether waiting forever is not itself an observation. In other words, should it not be possible to state in the semantics that a given program terminates or doesn't terminate for a given initial condition? This point can be argued either way for programs (since knowing that a program terminates before infinity is not very practical information). However, for digital systems (or reactive systems in general), it is clear that simple before and after conditions are not a sufficient theory. First of all, termination for these systems is not well defined, and moreover the meaning of what these systems are supposed to *do* is inseparably linked with the evolution of events in time [Pnu77].[1] What we need is a formal theory in which we can reason about temporal aspects of a system's behavior.

---

[1] The question of termination is in any event not undecidable for hardware systems, since they are not computationally universal (only programming languages are).

2.1 TEMPORAL LOGIC

Temporal logic (or *tense* logic) is a system devised by philosophers expressly for making statements about changes in time [Bur84]. In temporal logic, the formula Fq is true in the present if q is true at some moment in the future. Similarly Pq is true in the present if q is true at some moment in the past. These tense operators, F and P, have duals which are generally given their own names. The formula Gq is equivalent to $\neg F \neg q$, meaning that q is true at every moment in the future. The formula Hq is equivalent to $\neg P \neg q$, meaning that q is true at every moment in the past. These operators can give surprisingly concise expressions of sentences with complex tense structures. For example, $q \Rightarrow FPq$ can be interpreted as "if q holds in the present, then at some time in the future q will have held in the past".

Whenever we deal with a formal logic, we would like to have a *model theory* for the logic, which tells us the set of imaginary universes in which a given formula is true. This attaches an intuitive meaning or *semantics* to formulas, which allows us to judge whether our axioms and inference rules are reasonable. The usual model theoretic semantics given to temporal logic (and modal logics in general) is the so-called *possible worlds* semantics. A *frame* in this semantics consists of a class S of states through which the system evolves, and a relation $<$ representing temporal order. Intuitively, given two states s and t, $s < t$ means that s occurs before t. A *model* is a frame with a *valuation* L, which assigns truth or falsehood to every atomic proposition (propositional letter) in every state.[2] The truth or falsehood of temporal formulas is relative to the present state. For example, the formula Fq is true in state s iff there exists a state t such that p is true in state t and $s < t$. Similarly, Pq is true in state s iff there exists a state t such that p is true in state t and $t < s$. Notice that a temporal formula acts like an open sentence, with one free parameter s representing the present state. Thus it defines a class of states in which the formula is true. Similarly, a state defines a class of formulas which are true in that state.

Given the above model theoretic semantics, the choice of axioms in the logic effectively characterizes the temporal ordering relation $<$. For example, the following axioms (in addition to the propositional tautologies) exactly characterize those frames whose $<$ relation is a partial order (transitive and antisymmetric) [Bur84]:

$$G(p \Rightarrow q) \Rightarrow (Gp \Rightarrow Gq) \tag{2.1}$$

$$H(p \Rightarrow q) \Rightarrow (Hp \Rightarrow Hq) \tag{2.2}$$

---

[2] These are usually called Kripke frames and Kripke models, after one of the first mathematicians to give a model theoretic interpretation of modal logic.

$$p \Rightarrow GPp \qquad (2.3)$$

$$p \Rightarrow HFp \qquad (2.4)$$

One inference rule (in addition to *modus ponens*) is required: by *temporal generalization*, if α is provable, we infer that $G\alpha$ and $H\alpha$ (that is, a tautology must hold true at all times, or perhaps, the rules of sound inference do not change with time). By specializing this system slightly, we can obtain logics characterizing a variety of models of time, including linear time, discrete time, and branching (non-deterministic) time. All of these results can be found in [Bur84].

2.1.1 Linear time

We usually think of time as a linearly ordered set, measuring it either with the real numbers or the natural numbers. A frame is linear when the temporal order relation is total. That is, for all states s,t, either $s < t$, $s = t$, or $t < s$. The linear temporal frames can be characterized by simply adding the following two axioms to the basic set (they are time reversal duals):

$$(FPq) \Rightarrow (Pq \vee q \vee Fq) \qquad (2.5)$$

$$(PFq) \Rightarrow (Pq \vee q \vee Fq) \qquad (2.6)$$

Linear temporal logic is usually extended by the *until* operator and the *since* operator. Informally, $p \, U \, q$ states that p will hold at some moment in the future, until which time q will hold at all moments. Similarly, $p \, S \, q$ states that p held at some moment in the past, since which time q has held at all moments. More precisely, $p \, U \, q$ is true in state s if there is some state t such that $s < t$ and q is true in state t, and for all $s < u < t$, p is true in state u. Similarly, $p \, S \, q$ is true in state s if there is some state t such that $t < s$ and q is true in state t, and for all $t < u < s$, p is true in state u.

2.1.2 Discrete time

It is common in engineering to model time as a discrete sequence (measured by the integers). Discrete dynamics are commonly used, for example, in signal processing and synchronous digital systems. A discrete frame is one in which every state has an immediate successor and an immediate predecessor. The linear discrete frames can be characterized by adding the following two axioms to those for linear time logic:

$$p \wedge Hp \Rightarrow FHp \qquad (2.7)$$

Model checking

$$p \wedge Gp \Rightarrow PGp \qquad (2.8)$$

It is useful in a discrete linear temporal logic to define a *next time* temporal operator. The formula Xq is true in state s when there is an immediate successor of x in which q is true. A state t is an immediate successor of s if $s < t$ and there does not exist a state u such that $s < u < t$. Thus, Xq is exactly equivalent to false U q, so its addition does not increase the expressiveness of the logic.

2.1.3 Branching time

A branching frame is one in which the temporal order $<$ defines a tree which branches toward the future. Thus, every instant has a unique past, but an indeterminate future. This is an inherently non-deterministic model of time, and hence is well suited, for example, for defining a semantics of non-deterministic programs. A frame is tree ordered when for all states s, t, u, if $t < s$ and $u < s$ then $t < u$, $t = u$ or $t > u$. In other words, the past of every state is linearly ordered. The tree ordered frames can be characterized by simply dropping (2.6) from the axioms of linear time logic.

Though pure tense logic can exactly characterize the branching time frames, it leaves something to be desired in describing non-determinism, our intuitive interpretation of branching time. When we deal with non-deterministic automata or programs, we typically define the result of a computation in terms of possibility or inevitability. For instance, a non-deterministic finite automaton (NFA) accepts a string if it may *possibly* accept the string. Conversely, a non-deterministic program aborts only if it must *inevitably* abort[3]. These notions of inevitability and possibility are not represented in an ordinary tense logic. They can be incorporated, however, by combining notions from temporal logic and modal logic.

We would like to interpret the branching structure of time as meaning that each instant of time has many possible futures, and that as time evolves from present to future, these possibilities are reduced. Thus, in the past, there existed possible futures which are now precluded. This interpretation gives rise to notions of necessity (inevitability) and possibility in tense logic [Tho84]. We think of the truth or falsehood of tense formulas as being relative to a given branch of the tree ordered frame (one possible evolution of time into the future). A branch is defined as a maximal linearly ordered set of states. We will write

---
[3] In either case, the functionality might be implemented by backtracking.

$q[s, b]$ if q holds in state s in branch b. Thus, $Fq[s,b]$ iff there exists a state t in b such that $s < t$ and $q[t, b]$. Similarly, $Pq[s, b]$ iff there exists a state t in b such that $t < s$ and $q[t, b]$. The notion that q is *necessarily* true is represented by the formula Aq. We will say $Aq[s, b]$ iff for all branches b' containing s, $q[s, b']$.

The notion that q is *possibly* true is represented by the formula Eq. We will say $Eq[s, b]$ iff for some branch b' containing s, $q[s, b']$. The modal operators A and E provide a kind of second order quantification over maximal linearly ordered subsets.[4]

According to this semantics for modal branching time logic, there may be possibilities in the past that are foreclosed in the present. For example, $q \Rightarrow HAFq$ is not valid. The fact of q in the present does not imply the necessity of q in the past. Thus, modal branching time logic might be termed the logic of regret. The logic can also express useful semantic properties of non-deterministic programs [BAMP81]. For example, if q represents the fact of a program terminating, then inevitable termination is expressed by the formula AFq (necessarily in the future q). Possible termination is expressed by EFq (possibly in the future q). If the proposition p represents a correct output of the program, then (inevitable) partial correctness is expressed by the formula $AG(q \Rightarrow p)$ (necessarily invariantly, termination implies correctness). Note that Pq, APq and EPq are all logically equivalent, since the past of a state is the same for any branch. Also note that A and E are dual, since Aq is equivalent to $\neg E \neg q$.

The interpretation of formulas over maximal linearly ordered subsets allows us to characterize the branching frames in a different way. To the axioms of linear temporal logic, we add the following two:

$$Ep \Rightarrow HEFp \qquad (2.9)$$

$$PAGp \Rightarrow Ap \qquad (2.10)$$

2.2 THE TEMPORAL LOGIC CTL

We will now take a detailed look at the model theory of a very simple subset of modal tense logic defined by Clarke and Emerson [CE81b]. The logic is called CTL, which stands for *Computation Tree Logic*.[5] Our interest in this logic is

---

[4] Classically, the symbol □ is used to represent necessity, and ◇ is used to represent possibility. The symbols A and E are used here for consistency with [BAMP81].

[5] CTL is actually a subset of a more general temporal logic described in [CE81a], adopting the syntax of [BAMP81].

that its operators have simple characterizations as fixed points that allow us to compute efficiently whether a formula is satisfied in a given state in a given finite model. This calculation is termed *model checking*.

In CTL, temporal operators occur only in pairs consisting of A or E, followed by F, G, U or X. Thus, past time operators are not allowed, and tense operators cannot be combined directly with the propositional connectives. To be exact:

1. Every atomic proposition is a CTL formula.
2. If f and g are CTL formulas, then so are

$$\neg f, \ (f \wedge g), \ AXf, \ EXf, \ A(fUg), \ E(fUg)$$

The remaining operators are viewed as being derived from these according to the following rules:

$$\begin{aligned} f \vee g &= \neg(\neg f \wedge \neg g) \\ AFg &= A(\text{true } U \ g) \\ EFg &= E(\text{true } U \ g) \\ AGf &= \neg E(\text{true } U \ \neg f) \\ EGf &= \neg A(\text{true } U \ \neg f) \end{aligned}$$

Note that since all operatorss are prefixed by A or E, the truth or falsehood of a formula depends only on the given state s, and not on the particular branch b.

As a means of verifying systems, we would like to determine the truth or falsehood of a formula in this logic with respect to a given state in a given finite model. This means, however, that we must be able to represent a reactive system, with non-terminating behaviors, using a finite model. To do this, we introduce a nonstandard model. This model is a triple (S, R, L), where S is the set of states, R is the *transition relation* and L is the valuation. The transition relation is the set of all pairs (s,t) such that t is an *immediate successor* of s. A standard branching model (*a.k.a.* computation tree) can be obtained by starting at a designated state s and unwinding the graph (S, R) into an infinite tree (provided every state has at least one successor). We can give a semantics for CTL formulas with respect to the nonstandard model which is equivalent to the standard semantics with respect to the corresponding infinite tree.[6]

---

[6] With one additional distinction: in CTL, the future is taken to include the present. Thus, if p holds in the present, then so does Fp.

A *path* of a model $K = (S, R, L)$ is an infinite sequence of states $(s_0, s_1, s_2 \ldots) \in S^\omega$ such that each successive pair of states (s_i, s_{i+1}) is an element of R. Every path is maximal linearly ordered subset of the branching model unwound from s_0.

The notation $K, s \models f$ means that the formula f is true in state s of model K. In the sequel, where the model is unambiguous, we will write simply $s \models f$. The semantics of CTL formulas with respect to a model (S, R, L) is given below:

$$\begin{aligned}
s \models p & \quad \text{iff} \quad s \in L(p), \text{ where } p \text{ is an atomic proposition} \\
s \models \neg f & \quad \text{iff} \quad s \not\models f \\
s \models f \wedge g & \quad \text{iff} \quad s \models f \text{ and } s \models g \\
s_0 \models AXf & \quad \text{iff} \quad \text{for all paths } (s_0, s_1, \ldots), s_1 \models f \\
s_0 \models EXf & \quad \text{iff} \quad \text{for some path } (s_0, s_1, \ldots), s_1 \models f \\
s_0 \models A(f \ U \ g) & \quad \text{iff} \quad \text{for all paths } (s_0, s_1, \ldots), \text{ for some } i, \\
& \qquad s_i \models g \text{ and} \\
& \qquad \text{for all } j < i, s_j \models f \\
s_0 \models E(f \ U \ g) & \quad \text{iff} \quad \text{for some path } (s_0, s_1, \ldots), \text{ for some } i, \\
& \qquad s_i \models g \text{ and} \\
& \qquad \text{for all } j < i, s_j \models f
\end{aligned}$$

2.3 FIXED POINTS

Emerson and Clarke [CE81a] showed that various branching time properties of programs can be characterized as fixed points of appropriate monotonic functionals. Later, they introduced the logic CTL, and showed that its operators can be characterized in this way [CE81b]. This characterization led to an efficient algorithm for the model checking problem – determining whether a given CTL formula is satisfied in a given state of a finite Kripke model.

To obtain the fixed point characterization, we detour briefly into Boolean algebra. Let us identify every logical formula with a subset of set S. The logical constant false is identified with the empty set, while true is identified with S. We identify $p \wedge q$ with the intersection of p and q, $p \vee q$ with the union of p and q, and $\neg p$ with the complement of p in S. It is not difficult to show that this representation of formulas satisfies the axioms of Boolean algebra. In fact, any representation of formulas satisfying the Boolean axioms is isomorphic to this one, for some S.

A *functional* is denoted $\lambda y.f$, where f is a formula, and y is a variable. The

Model checking

variable y acts as a place holder. When applied to a parameter p, the functional $\lambda y.f$ yields f with p substituted for y. For example, if $\tau = \lambda y.(x \wedge y)$, then $\tau(\text{true}) = x \wedge \text{true} = x$.

A *fixed point* of a functional τ is any p such that $\tau(p) = p$. For example, if $\tau = \lambda y.(x \wedge y)$, then $x \wedge y$ is a fixed point of τ, since $\tau(x \wedge y) = x \wedge (x \wedge y) = x \wedge y$.

A functional τ is *monotonic* when $p \subseteq q$ implies $\tau(p) \subseteq \tau(q)$. For example, $\lambda y.(x \wedge y)$ is monotonic, since $p \subseteq q$ implies $x \wedge p \subseteq x \wedge q$. On the other hand, $\lambda y.(x \wedge \neg y)$ is is not monotonic, since false $\subseteq$ true, but $x \wedge \text{true} \not\subseteq x \wedge \text{false}$.

Fixed points of monotonic functionals have some useful properties. Tarski showed that a monotonic functional has a *least fixed point*, which is the intersection of all the fixed points [Tar55]. It also has a *greatest fixed point*, which is the union of all the fixed points. The least and greatest fixed points of $\lambda y.f$ are denoted $\mu y.f$ and $\nu y.f$ respectively.

Both extremal fixed points can be characterized as the limit of a series obtained by iterating the functional, provided the functional is *continuous*. A functional τ is union-continuous when for any non-decreasing infinite sequence of sets $p_1 \subseteq p_2 \subseteq p_3 \subseteq \cdots$, $\cup_i \tau(p_1) = \tau(\cup_i p_i)$. In other words τ applied to the union of the sequence is equal to the union of τ applied to each element p_i of the sequence. Tarski showed that if τ is monotonic and union-continuous, then the least fixed point of τ is $\cup_i \tau^i(\text{false})$, the union of the sequence obtained by iterating τ with the initial value false.

This result is not difficult to explain: Because τ is monotonic, the sequence obtained by starting with false (the empty set) and iterating τ must be non-decreasing. Further, every element of this sequence is contained in the least fixed point of τ. This is shown by supposing that $\tau^i(\text{false}) \subseteq \mu y.\tau(y)$. It follows that $\tau^{i+1}(\text{false}) \subseteq \tau(\mu y.\tau(y)) = \mu y.\tau(y)$. Hence, by induction $\tau^i(\text{false}) \subseteq \mu y.\tau(y)$ for all i. That the union of the sequence is a fixed point follows from union-continuity: $\tau(\cup_i \tau^i(\text{false})) = \cup_i \tau(\tau^i(\text{false})) = \text{false} \cup_i \tau^{i+1}(\text{false}) = \cup_i \tau^i(\text{false})$. Hence, the union of the sequence is the least fixed point.

There is a similar result for the greatest fixed point: A functional τ is said to be intersection-continuous when for any non-decreasing infinite sequence of sets $p_1 \subseteq p_2 \subseteq p_3 \subseteq \cdots$, $\cap_i \tau(p_1) = \tau(\cap_i p_i)$. If τ is monotonic and intersection-continuous, then the greatest fixed point of τ is $\cap_i \tau^i(\text{true})$, the intersection of the sequence obtained by iterating τ with the initial value true.

We note that if the state set S is finite, then any monotonic τ is necessarily union-continuous and intersection-continuous. This is because every sequence of subsets $p_1 \subseteq p_2 \subseteq \cdots$ of a finite set S must have a maximum element p_m where $p_m = \cup_i p_i$. If τ is monotonic then $\cup_i \tau(p_i) = \tau(p_m)$, hence $\cup_i \tau(p_i) = \tau(\cup_i p_i)$. We can show similarly that $\cap_i \tau(p_i) = \tau(\cap_i p_i)$.

2.4 CTL MODEL CHECKING

Now let us identify every CTL formula f with $\{s \mid s \models f\}$, the set of states in which the formula is true. We note that EFp is logically equivalent to $p \vee EX\ EFp$. That is, EFp is true in the current state s when p is true in s or EFp is true in some successor of s. Two logically equivalent formulas are satisfied in the same set of states. Thus $EFp = p \vee EX\ EFp$. This makes EF a fixed point of the functional $\tau = \lambda y.p \vee EXy$. This function is easily shown to be monotonic. In fact, it satisfies a stronger condition of linearity with respect to union: $\tau(x \cup y) = \tau(x) \cup \tau(y)$.

The functional $\tau = \lambda y.p \vee EXy$ has a least fixed point which we can show is exactly EFp. This can be seen by supposing that y is a fixed point of τ. It follows that $p \subseteq y$ and $EXy \subseteq y$. From the latter, we infer that y is closed under R^{-1}, that is, y contains all of its predecessors. Hence, no path beginning in a state not in y reaches a state in y. Since $p \subseteq y$, it follows that $EFp \subseteq y$. Thus, EFp is contained in all fixed points of τ, making it the least fixed point of τ.

Similar fixed point characterizations can be obtained for the other CTL operators. In particular,

Theorem 1 (Clarke-Emerson) *For all models* (S, R, L),

1. $EFp = \mu y.(p \vee EXy)$
2. $EGp = \nu y.(p \wedge EXy)$
3. $E(q\ U\ p) = \mu y.(p \vee (q \wedge EXy))$

Since in the case of model checking we are dealing only with models with finite state set S, each of the above fixed points can be characterized as the limit of

Model checking

a series obtained by iterating the corresponding functional. That is, provided S is finite: [7]

$$EFp = \cup_i (\lambda y.(p \vee EXy))^i (\text{false}) \quad (2.11)$$
$$EGp = \cap_i (\lambda y.(p \wedge EXy))^i (\text{true}) \quad (2.12)$$
$$E(p \; U \; q) = \cup_i (\lambda y.(p \vee (q \wedge EXy)))^i (\text{false}) \quad (2.13)$$

Since the set of states S is finite, a limit must occur in these series after a finite number of iterations. In fact, the maximum number of iterations required to reach the limit is $|S|$. This is because a monotonic functional applied to false must yield a non-decreasing sequence. There is no strictly increasing sequence of sets of length greater than $|S|$, (excluding the empty set, which is the result of 0 iterations). Thus, any non-decreasing sequence of length greater than $|S|$ must have a repeated element, which must be the limit. A monotonic functional applied to true yields a non-increasing sequence, with the same result. This gives us an effective procedure for evaluating the fixed points:

To compute the least {or greatest} fixed point of τ :
   let $Y = \text{false}$; {or $Y = \text{true}$}
   do
      let $Y' = Y$, $Y = \tau(Y)$
   until $Y' = Y$;
   return Y

As an example, consider computing EFp in the following Kripke model:[8]

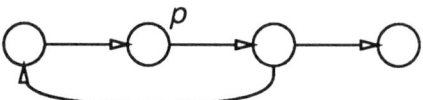

---

[7] In fact, this result does not generalize to infinite state models. The functionals whose least fixed points characterize EF and EU are union-continuous, hence their least fixed points are equal to the union of corresponding iteration sequence, even for infinite models. However, in the case of EG, we find that the functional $\tau = \lambda y.(p \wedge EXy)$ is not intersection continuous, and in fact one can contruct infinite state models in which EGp is false in a given state, whereas $\tau^i(\text{false})$ holds for all natural numbers i.

[8] We represent a Kripke model pictorially by drawing the graph (S, R) and labeling each state with the atomic propositions which are true in that state.

Since $|S| = 4$, the number of iterations required to produce the fixed point is at most 4. Therefore, let us compute $\tau^i(\text{false})$ for $i = 1\ldots 4$, where $\tau = \lambda y.p \vee EXy$. After the first iteration, we have $\tau^1(\text{false}) = p \vee EX\text{false} = p$:

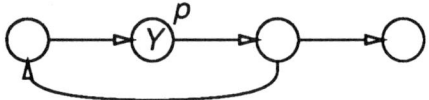

After the second iteration, we have $\tau^2(\text{false}) = p \vee EXp$:

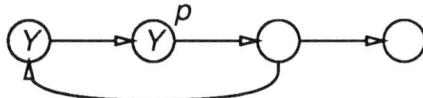

After the third iteration, we have $\tau^3(\text{false}) = p \vee EX(p \vee EXp)$:

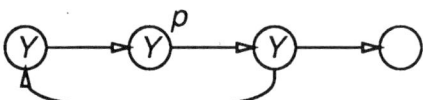

which is a fixed point, since the next iteration, $\tau^4[\text{false}]$ produces the same result. Notice that at each iteration i, we have the set of states s_0 such that there exists a path $(s_0, s_1, s_2, \ldots)$ where p is true at some state less than i. In effect, each iteration propagates the formula EFp backward in the graph by one step. When this process reaches a fixed point, we have labeled exactly the set of states on a path to a state labeled with p.

As a second example, consider computing EGp in the following Kripke model:

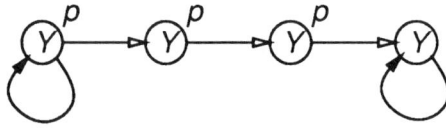

After the first iteration, we have $\tau^1(\text{true}) = p \wedge EX\text{true} = p$:

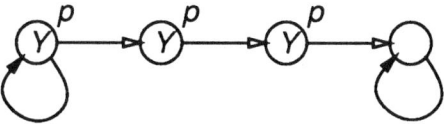

After the second iteration, we have $\tau^2(\text{true}) = p \wedge EXp$:

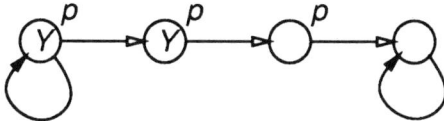

After the third iteration, we have $\tau^3(\text{true}) = p \wedge EX(p \wedge EXp)$:

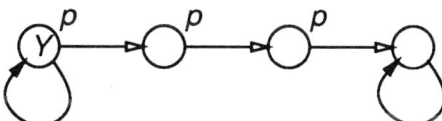

This is the greatest fixed point, since the next iteration $\tau^4(\text{true})$ produces the same result. Notice that at iteration i, we have the set of states such that there exists a path of length i where every state satisfies p. When we reach a fixed point, every state in the set has a successor in the set satisfying p, hence for every state in the set, there exists an infinite path where p is always true.

The operators EX, $E(\ U\ )$ and EG are actually sufficient to characterize the entire logic, since the remaining operators can be derived from these three according to the following rules:

$$\begin{aligned}
EFp &= E(\text{true } U\ p) \\
AXp &= \neg EX \neg p \\
AGp &= \neg EF \neg p \\
A(q\ U\ p) &= \neg(E(\neg p\ U\ \neg q \wedge \neg p) \vee EG \neg p)
\end{aligned}$$

For this reason, in the sequel, we will consider only the operators EX, $E(\ U\ )$ and EG. However, for completeness, here are the fixed point characterizations of the remaining operators:

$$AGp\ =\ \nu Y.(p \wedge AXY)$$
$$A(q\ U\ p)\ =\ \mu Y.p \vee (q \wedge AXY)$$

The fixed point characterization provides an effective algorithm for the model checking problem. In fact, a more efficient algorithm exists, based on breadth first search and the calculation of strongly connected components in the graph (S, R) [CES86]. Both of these algorithms suffer from the state explosion problem, however; it is necessary to construct the complete state graph of the system being modeled before model checking can be applied. Since the number of states of a system grows exponentially in the number of its components, these algorithms can only be applied to small systems.

3

SYMBOLIC MODEL CHECKING

In the previous chapter, we equated a CTL formula with the set of states in which the formula is true. We showed how the CTL operators can thus be characterized as fixed points of certain continuous functionals in the lattice of subsets, and how these fixed points can be computed iteratively. This provides us with a model checking algorithm for CTL, but requires us to build a finite Kripke model for our system and hence leads us to the state explosion problem. In this chapter, we will explore a method of model checking that avoids the state explosion problem in some cases by representing the Kripke model implicitly with a Boolean formula. This allows the CTL model checking algorithm to be implemented using well developed automatic techniques for manipulating Boolean formulas. Since the Kripke model is symbolically represented, there is no need to actually construct it as an explicit data structure. Hence, the state explosion problem can be avoided.

3.1 BOOLEAN REPRESENTATIONS

Our first step is to develop a way to equate Boolean formulas with sets of states and relations on states. To do this, let $\mathcal{S}$ be a set of variables, and consider the Boolean algebra $2^{2^{\mathcal{S}}}$. As before, we identify each formula with a set, in this case a set of subsets of $\mathcal{S}$. We think of each subset of $\mathcal{S}$ as representing a truth valuation, where variables in the set are true, and variables not in the set are false. Thus, each variable $v \in \mathcal{S}$ is identified with the set of sets containing v, $\{a \in 2^{\mathcal{S}} \mid v \in a\}$.

Now, let our set of states S be the set of Boolean vectors $\{\text{true}, \text{false}\}^n$. In this way, we can identify a functional with a set of states. Consider, for example, the vector functional $\lambda(v_1,\ldots,v_n).f$. When applied to a vector $(a_1,\ldots,a_n)$, this functional yields f with v_i substituted for a_i. Vector functionals obey the usual axioms associated with λ:

$$(\lambda\bar{v}.v_i)(\bar{a}) = a_i \tag{3.1}$$

$$(\lambda\bar{v}.f)(\bar{a}) = f \text{ if } f \text{ does not contain any } v_i \tag{3.2}$$

$$(\lambda\bar{v}.f(g))(\bar{a}) = f((\lambda\bar{v}.g)(\bar{a})) \tag{3.3}$$

Now suppose we have a vector functional $\tau = \lambda(v_1,\ldots,v_n).f$ that is closed, in the sense that it takes every Boolean vector $a \in \{\text{true},\text{false}\}^n$ to either true or false. A formula f that contains only variables $v_1,\ldots,v_n$ yields such a functional. In this case the functional τ uniquely characterizes a set of Boolean vectors – exactly those vectors (states) a such that $\tau(a) = \text{true}$. For any set of states p, we have a unique representation $\mathbf{p}$ such that

$$p = \lambda(v_1,\ldots,v_n).\mathbf{p} \tag{3.4}$$

Note that p is a set of Boolean vectors, while $\mathbf{p}$ is a set of subsets of $\mathcal{S}$. The representation is fixed by the particular choice of variables $v_1,\ldots,v_n$.

Lest this seem a useless exercise, let us move on to representations of relations. Consider the vector-pair functional $\tau = \lambda((v_1,\ldots,v_n),(v'_1,\ldots,v'_n)).f$ which takes a pair of vectors $(v_1,\ldots,v_n)$ and $(v'_1,\ldots,v'_n)$ as argument. If this functional is closed, it characterizes a set of pairs of states (a,a') such that $\tau(a,a') = \text{true}$. Thus, for any relation R on states we have a unique representation $\mathbf{R}$ such that

$$R = \lambda((v_1,\ldots,v_n),(v'_1,\ldots,v'_n)).\mathbf{R} \tag{3.5}$$

Note that R is a set of pairs of vectors, while $\mathbf{R}$ is a set of subsets of $\mathcal{S}$. The representation is fixed by the particular choice of variables $v_1,\ldots,v_n$ and $v'_1,\ldots,v'_n$. Thus, each set or relation has a unique representation in 2^{2^S}, though the same Boolean formula might be use to represent either a set or a relation, depending on the context. Note, however, that $v_1 \wedge v'_1$ does not represent any set, since $\lambda(v_1,\ldots,v_n).(v_1 \wedge v'_1)$ is not closed.

3.2 SYMBOLIC MODELS

Now, suppose we have a model $K = (S, R, L)$ where the state set S is the set of Boolean vectors $\{\text{true},\text{false}\}^n$ and the transition relation R is characterized by

Symbolic model checking

equation (3.5). Let us postulate a vector of atomic propositions $\bar{l} = l_1, \ldots, l_n$ whose truth values correspond to the elements of the state vector. Thus, the labeling L maps atomic proposition l_i onto the set of vectors $(a_1, \ldots, a_n)$ such that $a_i = \text{true}$. In other words,

$$L(l_i) = \lambda(v_1, \ldots, v_n).v_i \qquad (3.6)$$

Thus, according to our symbolic representation of sets, $\mathbf{l}_i = v_i$. In the sequel, we will assume without loss of generality that $\bar{l} = \bar{v}$.

To characterize the CTL operators with respect to our model, we need to introduce the notion of Boolean quantification. By the formula $\exists (v_1, \ldots, v_n).f$, we mean that f is true for some Boolean vector $(v_1, \ldots, v_n)$. In other words,

$$\exists (v_1, \ldots, v_n).f = \bigvee_{\bar{a} \in \{\text{true}, \text{false}\}^n} (\lambda(v_1, \ldots, v_n).f)(a) \qquad (3.7)$$

Similarly,

$$\forall (v_1, \ldots, v_n).f = \bigwedge_{\bar{a} \in \{\text{true}, \text{false}\}^n} (\lambda(v_1, \ldots, v_n).f)(a) \qquad (3.8)$$

Now suppose that the set of states p is characterized by equation (3.4). The formula EFp identifies the set of states $(v_1, \ldots, v_n)$ such that there exists a state $(v'_1, \ldots, v'_n)$ such that the pair $((v_1, \ldots, v_n), (v'_1, \ldots, v'_n))$ is in R and $(v'_1, \ldots, v'_n)$ is in p. That is,

$$EXp = \lambda \bar{v}.\exists \bar{v}'.(R(\bar{v}, \bar{v}') \wedge p(\bar{v}')) \qquad (3.9)$$

where $\bar{v} = (v_1, \ldots, v_n)$ and $\bar{v}' = (v_1, \ldots, v_n)$. Now according to equation (3.5), $R(\bar{v}, \bar{v}') = \mathbf{R}$ and according to equation (3.4), $p(\bar{v}') = \mathbf{p}'$ where $\mathbf{p}'$ is obtained by substituting v'_i for v_i in $\mathbf{p}$. This gives us $EXp = \lambda \bar{v}.\exists \bar{v}'.(\mathbf{R} \wedge \mathbf{p}')$. Hence, according to our Boolean representation,

$$EX\mathbf{p} = \exists \bar{v}'.(\mathbf{R} \wedge \mathbf{p}') \qquad (3.10)$$

In other words, given the representation of p, we can find the representation of EXp by changing every variable v_i to v'_i, taking the conjunction with $\mathbf{R}$, then quantifying out the variables $v_1, \ldots, v_n$ existentially.

As an example, consider a model in which $\bar{v} = (b)$, $\bar{v}' = (b')$, and $\mathbf{R} = b \vee b'$. The Kripke model (S, R, L) is depicted below:

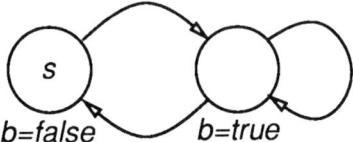

Let's say we want to determine the set of states satisfying $EX\neg b$. According to equation (3.10):

$$\begin{aligned}
EX\neg \mathbf{b} &= \exists (b').(\mathbf{R} \wedge (\neg b)') \\
&= \exists (b').((b \vee b') \wedge (\neg b')) \\
&= \exists (b').(b \wedge \neg b') \\
&= (b \wedge \neg \text{false}) \vee (b \wedge \neg \text{true}) \\
&= b
\end{aligned}$$

By a few Boolean algebra manipulations, we arrive at a formula exactly characterizing the states satisfying EXb. To find the truth value in any particular state, we can simply plug in the components of the state vector for the appropriate variables. For example, the truth value of EXp in state s is $(\lambda(b).b)(\text{false}) = \text{false}$.

As a more elaborate example, consider a model of a two bit binary counter circuit, as depicted in figure 3.1. Its transition relation is characterized by

$$\mathbf{R} = (v_0' \iff \neg v_0) \wedge (v_1' \iff v_0 \oplus v_1)$$

where $\bar{v} = (v_1, v_0)$ and $\bar{v}' = (v_1', v_0')$. Let us calculate symbolically the set of states satisfying $EX(v_0 \wedge v_1)$. According to equation (3.10):

$$\begin{aligned}
EX\neg \mathbf{v_0} \wedge \mathbf{v_1} &= \exists (v_0', v_1').(\mathbf{R} \wedge (v_0 \wedge v_1)') \\
&= \exists (v_0', v_1').(\neg v_0 \wedge v_1 \wedge v_0' \wedge v_1') \\
&= \exists (v_1').(\neg v_0 \wedge v_1 \wedge \text{false} \wedge v_1' \vee \neg v_0 \wedge v_1 \wedge \text{true} \wedge v_1') \\
&= (\neg v_0 \wedge v_1 \wedge \text{false}) \vee (\neg v_0 \wedge v_1 \wedge \text{true}) \\
&= \neg v_0 \wedge v_1
\end{aligned}$$

It should be no surprise that the state $(1,1)$ is preceded only by the state $(1,0)$, which is exactly the result we obtain. Note that a fair amount of Boolean algebra manipulation went into the first step. This computation might be carried out automatically by expanding all formulas into minterm canonical form, or a suitable cube covering, although much more efficient methods exist.

Symbolic model checking

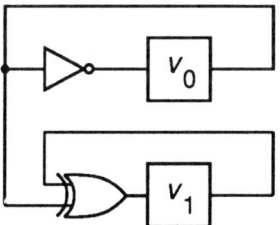

Figure 3.1 Two bit binary counter circuit

We postpone questions of automating the Boolean algebra to the next section, however.

For the moment, let us consider the remaining CTL operators, which have fixed point characterizations in terms of EX. For example, suppose in our simple example, where $\mathbf{R} = b \vee b'$ that we would like to compute the set of states satisfying EFb. This formula is the least fixed point of $\tau = \lambda y. b \vee EXy$, which we can obtain as the limit of the series $\cup_i \tau^i(\text{false})$. Evaluating the first few iterations of the series in our Boolean representation, we have:

$$\begin{aligned}
\tau^1[\text{false}] &= b \vee EX \text{false} \\
&= b \\
\tau^2[\text{false}] &= b \vee EXb \\
&= b \vee \exists b'.((b \vee b') \wedge b') \\
&= \text{true} \\
\tau^3[\text{false}] &= b \vee EX \text{true} \\
&= \text{true}
\end{aligned}$$

A fixed point is reached after two iterations. The set of states satisfying EFb is $\lambda(b).\text{true} = S$.

In the case of our two bit counter, evaluating $EF(v_1 \wedge v_0)$, we get

$$\begin{aligned}
\tau^1[\text{false}] &= v_0 \wedge v_1 \\
\tau^2[\text{false}] &= v_1 \\
\tau^3[\text{false}] &= v_0 \vee v_1 \\
\tau^4[\text{false}] &= \text{true}
\end{aligned}$$

The state sets characterized by these formulas are exactly those we would obtain using the fixed point model checking procedure on an explicit model, in effect propagating the formula $EF(v_1 \wedge v_0)$ backward in the state graph.

function eval(*f*)
    case
        f an atomic proposition: **return** *f*
        f = ¬*p*: **return** ¬eval(*p*)
        f = *p* ∨ *q*: **return** eval(*p*) ∨ eval(*q*)
        f = *EXp*: **return** evalEX(eval(*p*))
        f = *E*(*p U q*): **return** evalEU(eval(*p*),eval(*q*),false)
        f = *EGp*: **return** evalEG(eval(*p*),true)
    end case
end function

function evalEX(**p**) = $\exists \bar{v}'.(\mathbf{R} \wedge \mathbf{p}')$

function evalEU(**p, q, y**)
    y′ = **q** ∨ (**p** ∧ evalEX(**y**))
    if y′ = **y then return y**
    else return evalEU(**p, q, y**′)
end function

function evalEG(**p, y**)
    y′ = (**p** ∧ evalEX(**y**))
    if y′ = **y then return y**
    else return evalEG(**p, y**′)
end function

Figure 3.2 Symbolic CTL model checking algorithm

The complete procedure for symbolic model checking is shown in figure 3.2. Note that to determine when a limit is reached in the fixed point series, it is necessary to determine when two symbolic representations of sets are equal. This makes it highly advantageous to use a canonical form. Also note that since the model has 2^n states, the maximum number of iterations needed to reach a fixed point is $2^n + 1$. In practice, however, the number of iterations required to reach a fixed point can be quite small, even for fairly complex systems.

3.3 BINARY DECISION DIAGRAMS

It should be clear that to make the symbolic model checking technique practical, we need an efficient automated method for manipulating Boolean formulas. In particular, we must be able to reduce formulas including existential quantifiers to a suitable representation, so that we can test formulas for equality. Simply rewriting formulas into minterm canonical form is inadequate, because of the exponential explosion of minterms. A more sophisticated approach might attempt to find an optimal "cube covering" for the formula (a sum-of-products representation). We would then lose the advantage of a canonical form, however. In any event, the most promising representation for our purposes is clearly the *Ordered Binary Decision Diagram* form developed by Bryant [Bry86]. The Ordered Binary Decision Diagram (OBDD) representation is canonical, and the conjunction and disjunction of two OBDD's can be computed in quadratic time. The representation has also been shown to be compact for a wide variety of functions occurring in digital circuits, although there are a few important examples which do not have compact OBDD representations (in particular, multipliers and barrel shifters). OBDD's have been used with considerable success for the comparison of switching functions [BBB+87, FB89]. We will simply carry over the same technology for the comparison of fixed point iterations.

OBDD's can represent Boolean functions (as well as sets of Boolean vectors, sets of vectors pairs, *etc.*) in the same way as ordinary Boolean formulas, by associating elements of the input vector with variables from a set S. In the case of OBDD's a canonical representation is obtained by imposing a total order on the set of variables S. The OBDD canonical representation for a Boolean function can then be obtained by reducing a related structure called an ordered decision tree. The truth value of an ordered decision tree is obtained by descending the tree from the root to a leaf. At each node along the path, one descends to the left child if the value of the variable labeling the node is false, and to the right child if the value is true. Each leaf of the tree is labeled with a value true or false which gives the result. The tree must respect the order on S, in the sense that the variables always occur in increasing order along any path from root to leaf. Reading the leaves from left to right, one obtains the truth table of the function being represented.

As an example, an ordered decision tree for $a \wedge b \vee c \wedge d$ is depicted in figure 3.3. Note that in this figure and in the sequel, we will use 0 to denote false and 1 to denote true when it is convenient.

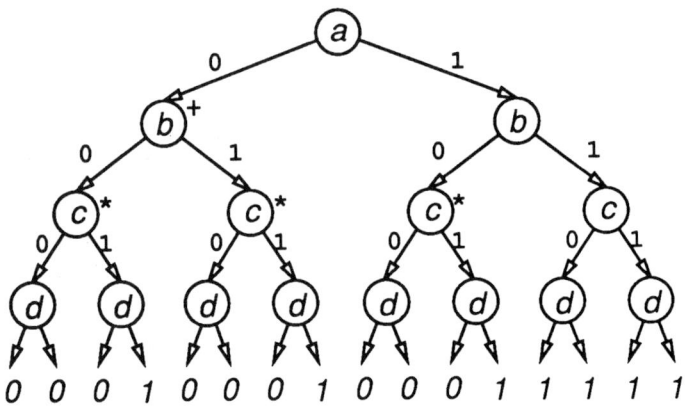

Figure 3.3 Ordered Decision Tree

The canonical OBDD form is a directed acyclic graph which can be obtained from the ordered decision tree by the following two steps:

1. Combine any isomorphic subtrees into a single tree.

2. Eliminate any nodes whose left and right children are isomorphic.

Steps 1 and 2 can be applied in a bottom up fashion, to yield the canonical OBDD representation in linear time. Bryant called this operation *Reduce*. The size of the resulting graph is strongly dependent on the order of the variables. This variable ordering, however, is the key to obtaining the canonical reduced form. This is what distinguishes OBDDs from the more general class of Binary Decision Diagrams described by Akers [Ake78].

As an illustration of reduction to canonical form, consider the the ordered decision tree of figure 3.3. The three nodes marked "*" are roots of isomorphic subtrees. Thus, they can be combined into a single subtree. In addition, from the node marked "+", one arrives at the same subtree when descending to the left or right (*ie.*, independently of the value of b), hence this vertex does not affect the value of the function and may be eliminated. The result of applying the *Reduce* operation to the tree of of figure 3.3 is depicted in depicted in figure 3.4. Note the significant reduction in the number of vertices, resulting essentially from redundancy in the truth table of the function.

Symbolic model checking

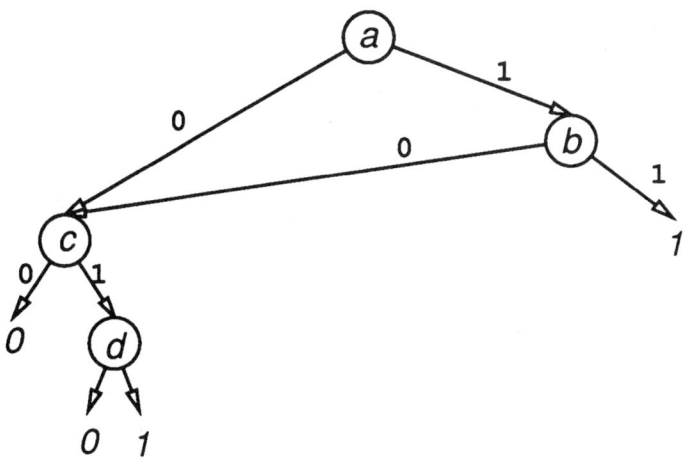

Figure 3.4 Ordered Binary Decision Diagram

Combining isomorphic subtrees makes OBDD's directed acyclic graphs rather than trees. We will call the leaves of these graphs *terminals*. Each internal node is a triple (v, l, h), where $v \in S$ is the variable labeling the node, l is the node on the left (low) branch and h is the node on the right (high) branch. The order on variables induces a pre-order on nodes:

Definition 1 *For all nodes* $n_1 = (v_1, l_1, h_1)$ *and* $n_2 = (v_2, l_2, h_2)$, $n_1 < n_2$ *when* $v_1 < v_2$. *For all nodes* $n = (v, l, h)$, $n < 0$ *and* $n < 1$.

We can define by induction a set of nodes that canonically represent the elements of the Boolean algebra 2^{2^S}.

Definition 2 *Let S be a totally ordered set. $OBDD(S)$ is the least set such that:*

1. $\{0, 1\} \subseteq OBDD(S)$ *and*

2. *For all nodes $l \neq h$ in $OBDD(S)$, for all $v \in S$, if $(v, l, h) < l$ and $(v, l, h) < h$ then $(v, l, h) \in OBDD(S)$.*

Having defined the set of OBDD nodes, we can associate each node n with a Boolean formula f_n. To do this, it is most convenient to define a Boolean operator $p \to (x, y)$, which is equivalent to x when p is true and y when p is false. The formula $p \to (x, y)$ is simply equivalent to $(p \land x) \lor (\neg p \land y)$.[1]

Definition 3 1. $f_0 = \text{true}$

2. $f_1 = \text{false}$

3. If $n = (v, l, h)$ is in $OBDD(\mathcal{S})$ then

$$f_n = v \to (f_h, f_l)$$

Starting from this definition, it is straightforward to prove that the OBDD nodes canonically represent our Boolean algebra. It is useful to start with a lemma stating that $f_{(v,l,h)}$ does not depend on any variable less than v:

Lemma 1 For all nodes $n = (v, l, h)$ in $OBDD(\mathcal{S})$, for all variables $v' \in \mathcal{S}$ such that $v' < v$,

$$(\lambda v'.f_n)(\text{false}) = (\lambda v'.f_n)(\text{true})$$

Proof. By structural induction. The terminal cases, $p = 0$ and $p = 1$ are trivial. We now assume theorem holds for l and h. By definition, $f_n = v \to (f_h, f_l)$. Thus, since $v' \neq v$:

$$\begin{aligned}
(\lambda v'.f_n)(\text{false}) &= (\lambda v'.v)(\text{false}) \to ((\lambda v'.(\lambda v'.f_h)(\text{false}), (\lambda v'.f_l)(\text{false})) \\
&= v \to (\lambda v'.(\lambda v'.f_h)(\text{false}), (\lambda v'.f_l)(\text{false})) \\
&= v \to (\lambda v'.(\lambda v'.f_h)(\text{true}), (\lambda v'.f_l)(\text{true})) \\
&= (\lambda v'.v)(\text{true}) \to (\lambda v'.(\lambda v'.f_h)(\text{true}), (\lambda v'.f_l)(\text{true})) \\
&= (\lambda v'.f_n)(\text{true})
\end{aligned}$$

□

---

[1] There is an alternative formulation of OBDDs due to Clarke [KC90] which does not require the children l and h of a node to be distinct, but requires that the variable labeling l and h be the successor of the variable labeling the parent. In this case, the OBDD for a formula f is exactly the minimal DFA recognizing the language $\lambda \mathcal{S}.f$. Thinking of OBDDs as minimal DFAs can provide useful insights into the complexity of representing certain classes of functions as OBDDs.

Symbolic model checking

We first show that there are no two distinct OBDD nodes representing the same formula, and second, that every Boolean formula is represented by some OBDD. The following theorem is essentially due to Bryant [Bry86], although the formalization is different, and as a result, it is hoped, the proof is somewhat simpler.

Theorem 2 (Bryant) *If $n, n' \in OBDD(\mathcal{S})$, then $f_n = f_{n'}$ implies $n = n'$.*

Proof. By induction over the pre-order on nodes. This order is well founded on the high end, since the terminals 0 and 1 are maximal. Thus, we assume that the statement of the theorem holds for nodes m and m' where $m > n$ and $m' > n'$.

Consider first the case where $n \sim n'$ (ie., neither $n > n'$ nor $n < n'$). Either p and q are both terminals, (in which case $n = n' = 0$ or $n = n' = 1$) or they are both non-terminals, $n = (v, l, h)$ and $n' = (v', l', h')$ where $v = v'$. For non-terminals, we have $v \to (f_h, f_l) = v \to (f_{h'}, f_{l'})$. This implies that $(\lambda v.f_h)(\text{true}) = (\lambda v.f_{h'})(\text{true})$ and $(\lambda v.f_l)(\text{false}) = (\lambda v.f_{l'})(\text{false})$. The lemma then gives us $f_h = f_{h'}$ and $f_l = f_{l'}$, hence by inductive hypothesis, $l = l'$ and $h = h'$. Thus $n = n'$.

Second, consider the case where $n < n'$. It follows that n is a non-terminal $n = (v, l, h)$. Further, by the previous lemma, $f_{n'}$ does not depend on v. Since $f_n = f_{n'}$, it follows that f_n does not depend on v. Since $f_n = v \to (f_h, f_l)$, it follows that $f_h = f_l$. By inductive hypothesis, then, $h = l$. This contradicts the definition of $OBDD(\mathcal{S})$, however, so this case is impossible.

A symmetric argument applies to the case $n' < n$. □

Theorem 3 *For any $f \in 2^{2^S}$, there exists $n \in OBDD(\mathcal{S})$ such that $f_n = f$.*

Proof. By induction on v, the least variable such that f depends on v. By inductive hypothesis, there exist nodes q and r in $OBDD(\mathcal{S})$ such that $f_q = (\lambda v.f)(\text{false})$ and $f_r = (\lambda v.f)(\text{true})$, since these depend only on variables greater than v. Further, q and r are distinct, since f depends on v. Thus, let $n = (v, q, r)$. □

Because two formulas that are equal are represented by the same OBDD node, testing two OBDDs for functional equality can be accomplished in constant time. This property of OBDDs is useful for determining when the limit of a fixed point series has been reached in the symbolic model checking algorithm. Since the OBDD nodes are in one-to-one correspondence with the elements of our Boolean algebra, we will simply drop the notation f_n in the sequence, and treat OBDD's exactly like Boolean formulas.

The Apply *algorithm*

Bryant describes an algorithm called *Apply*, which applies an arbitrary Boolean operation $f : \{0,1\}^2 \rightarrow \{0,1\}$ to two OBDDs. The operation f can be any of the 16 Boolean functions of two variables. We are interested, of course, in the operations $\vee$, $\wedge$ and $\neg$, but other operations such as exclusive-or are possible.

The *Apply* algorithm works by recursive descent on two OBDD's. It uses a hash table to store the result returned for each pair of nodes, so that the result for a given pair only has to be computed once. This is essentially dynamic programming. As a result, *Apply* has quadratic complexity.

To see how *Apply* works when given a pair of nodes (p, q), let us break the problem of computing $f(p, q)$ into cases. If p and q are terminals, then we simply return $f(p, q)$ according to the truth table for f. On the other hand, suppose p is an internal node (v, l, h) and $p < q$. In this case, q does not depend on v. Thus, we can show that $f(p, q) = v \rightarrow (f(h, q), f(l, q))$. We can call *Apply* recursively to compute compute $h' = f(h, q)$ and $l' = f(l, q)$. Then if $h' = l'$ we simply return h', otherwise we return an OBDD node (v, l', h'). We use a hash table to store the OBDD nodes, so that we never create more than one copy in memory of a given node.

The case $p > q$ is symmetric to the one above. In the remaining case, $p \sim q$, either both are terminals, or $p = (v_p, l_p, h_p)$ and $q = (v_q, l_q, h_q)$ where $v_p = v_q$. As a result, we can show that $f(p, q) = v \rightarrow (f(l_p, l_q), f(h_p, h_q))$. Thus, we call *Apply* recursively to compute $l' = f(l_p, l_q)$ and $h' = f(h_p, h_q)$. Again, if $h' = l'$ we simply return h', otherwise we return an OBDD node (v, l', h').

Since each subproblem can generate two more subproblems, it might seem that this algorithm is exponential. However, the number of node pairs we visit is bounded by $|p| \times |q|$, where $|p|$ is the number of nodes used in the representation of p. Since we store each result in a hash table, the number of times we have to carry out the above calculation is at most $|p| \times |q|$, making the algorithm quadratic. Bryant shows that this upper bound is tight, since there exist functions p and q for which the size of the result $f(p,q)$ is $|p| \times |q|$.

The Compose algorithm

Bryant also gives an algorithm called *Compose* which computes $(\lambda \bar{v}.p)(\bar{q})$, where $\bar{q}$ is a vector of OBDD's. The compose algorithm could easily be adapted to compute $\mathbf{p'}$ from $\mathbf{p}$, by letting $q_i = v'_i$. In general, however, we will assume that S is ordered so that if $v_i < v_j$, then $v'_i < v'_j$. This allows us to compute $\mathbf{p'}$ from $\mathbf{p}$ by simply substituting the node labelings, while preserving the OBDD structure.

Since $\exists(v).f = (\lambda v.f)(\text{false}) \vee (\lambda v.f)(\text{true})$, we could also adapt the *Compose* algorithm to computing $\exists \bar{v}.f$ for a vector of variables $\bar{v}$ by quantifying the variables one at a time. On the other hand, there is a procedure which is usually more eficient, and can compute $\exists \bar{v}.(f \wedge g)$ in a single recursive pass over f and g. We call this algorithm *AndExists*. In this algorithm, we quantify out variables immediately when the results of conjunction subproblems are obtained. This usually results in a substantial reduction in the size of the intermediate results by reducing the number of variables. The *AndExists* algorithm can be used to compute $EX\mathbf{p} = \exists \bar{v}'.(\mathbf{R} \wedge \mathbf{p'})$.

The AndExists algorithm

The *AndExists* algorithm takes as its arguments a vector $\bar{v}$ of variables and a pair of OBDD's (p,q), and returns $\exists \bar{v}.(p \wedge q)$. It is basically a modification of *Apply*, where $f(p,q) = p \wedge q$. In this algorithm, however, before we return a result, $r = (v, l, h)$, we test the variable v to see if it occurs in the vector $\bar{v}$. If it does, we call *Apply* to compute $l \vee h$, since $\exists v.(v \to (h,l)) = h \vee l$. Otherwise, we return r.

The motivation for this algorithm is to avoid producing the entire OBDD for $p \wedge q$, which has $2n$ variables, where n is the number of state variables of the model. This is done by applying existential quantification to the results of subproblems as soon as they become available, yielding a result with only n variables. Empirically, this provides a substantial savings in space.

As in the *Apply* algorithm, a hash table is used to avoid resolving previously computed subproblems. The maximum size of this table is $|p| \times |q|$. However, unl..e in the *Apply* algorithm, the recursive calls cannot be executed in constant

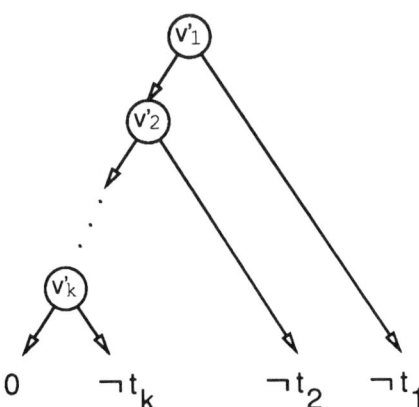

Figure 3.5 Variable ordering for 3-SAT reduction

time. This is because each call may require a ∨ operation to be performed. At present, the author is unaware of a bound on the complexity of *AndExists* better than $O(|p| \times |q| \times 2^{2n})$, which is simply the number of ∨ problems to be solved ($|p| \cdot |q|$ in the worst case) times the square of the largest possible OBDD size, 2^n. In practice, this number of operations has not been observed, so one might conjecture that there is a tighter bound. It seems unlikely that a polynomial bound will be found, however, since it is easily shown that if vector existential quantification on OBDDs can be computed in polynomial time, then P = NP.

The proof of this is by reduction from 3-SAT, as follows: Let $f = t_1 \wedge t_2 \wedge \cdots t_k$ be a 3-SAT formula, that is, $t_i = (x_i \vee y_i \vee z_i)$, where x_i, y_i and z_i are positive or negative literals. The OBDD representation of each t_i has no more than 3 non-terminals. Now introduce new variables $\bar{v}' = (v'_1, v'_2, \ldots, v'_k)$, corresponding to the terms of f, and let

$$f' = \bigvee_{1 \leq i \leq k} \left(\neg t_i \wedge v'_i \wedge \bigwedge_{1 \leq j < i} \neg v'_j \right)$$

For a suitable variable ordering, the OBDD representing f' has no more than $4k$ non-terminals (see figure 3.5), hence can be built in polynomial time. The formula f is satisfiable iff $\exists \bar{v}'. f' \neq$ true. Thus, if $\exists \bar{v}'. f'$ can be computed in polynomial time, then P = NP.

As an aside, it is not difficult (though a bit tedious) to show that the symbolic CTL model checking problem is PSPACE-complete. To show PSPACE-

Symbolic model checking

hardness, one starts with a polynomial space bounded Turing machine, introduces a sufficient number of Boolean variables to encode the entire tape, plus the pointer and the finite control, then expresses the transition relation of the entire system as a Boolean formula **R**. The acceptance condition for the machine can easily be expressed in CTL. To show that the problem is in PSPACE, one can show that the problem can be reduced to satisfiability of a QBF formula of polynomial size, using the "iterative squaring" technique of Burch, *et al.* [BCM+90]. Details are left to the reader.

3.4 EXAMPLES

Although the worst case complexity of symbolic model checking is high (using OBDDs or any other Boolean function representations), in practice the worst case complexity is rarely achieved, and the symbolic technique can in some cases be dramatically more efficient than previous methods. As an illustration of this, let's look at two hardware examples – a synchronous fair bus arbiter, and an asynchronous distributed mutual exclusion ring circuit (the one studied by David Dill in his thesis [Dil88] and designed by Alain Martin [Mar85]).

3.4.1 Synchronous state machines

For a synchronous finite state machine, the transition relation can be given as a conjunction of Boolean formulas, each determining the new state of one register as a function of its old state and the inputs. Let $\bar{v} = (v_1, v_2, \ldots, v_n)$ be a vector of Boolean variables representing the state of the registers in the circuit, and let $\bar{w} = (w_1, w_2, \ldots, w_m)$ be a vector of variables representing the values of the inputs to the circuit. For all $i = 1 \ldots n$, let f_i be a Boolean formula characterizing the truth value of register i in the next state. The transition relation of the state machine can be expressed as a Boolean formula in the following form:

$$\mathbf{R} = \bigwedge_{i=1}^{n} \mathbf{R}_i, \quad \text{where } \mathbf{R}_i = (v'_i \iff f_i). \tag{3.11}$$

In general, for models of synchronous systems, the transition relation is a conjunction of formulas representing the individual components of the system, since transitions of the components are *simultaneous*. The outputs of the state machine can be given as Boolean functions of the inputs and registers. These

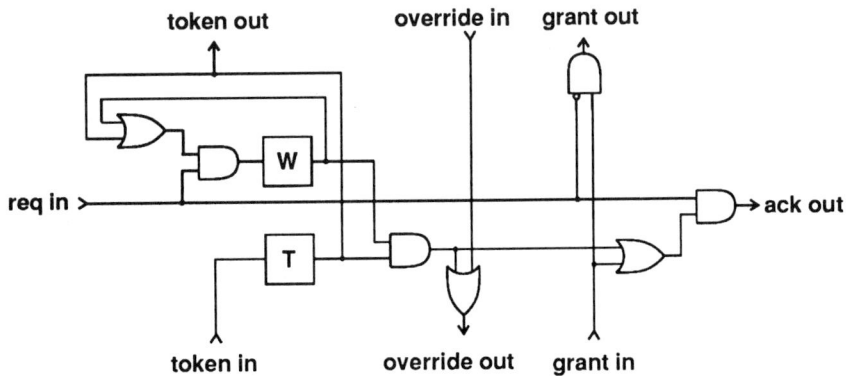

Figure 3.6 Cell of synchronous arbiter circuit

functions can be substituted for atomic propositions in CTL formulas, so there is no need to introduce variables to represent the outputs.

As an example of a synchronous state machine, we will consider a synchronous bus arbiter circuit. The purpose of the bus arbiter is to grant access on each clock cycle to a single client among a number of clients contending for the use of a bus (or other resource). The inputs to the circuit are a set of request signals $req_0 \ldots req_{k-1}$, and the outputs are a set of acknowledge signals $ack_0 \ldots ack_{k-1}$. Normally, the arbiter asserts the acknowledge signal of the requesting client with the lowest index. However, as requests become more frequent, the arbiter is designed to fall back on a round robin scheme, so that every requester is eventually acknowledged. This is done by circulating a token in a ring of arbiter cells, with one cell per client. The token moves once every clock cycle. If a given client's request persists for the time it takes for the token to make a complete circuit, that client is granted immediate access to the bus.

The basic cell of the arbiter is depicted in figure 3.4.1. This cell is repeated k times, as shown in figure 3.4.1. Each cell has a request input and an acknowledge output. The grant output of cell i is passed to cell $i+1$, and indicates that no clients of index less than or equal to i are requesting. Hence, a cell may assert its acknowledge output if its grant input is asserted. Each cell has a register T which stores a one when the token is present. The T registers form a circular shift register which shifts up one place each clock cycle. Each cell also has a register W (for "waiting") which is set to one when the request input is asserted and the token is present. The register remains set while the request persists, until the token returns. At this time, the cell's override and

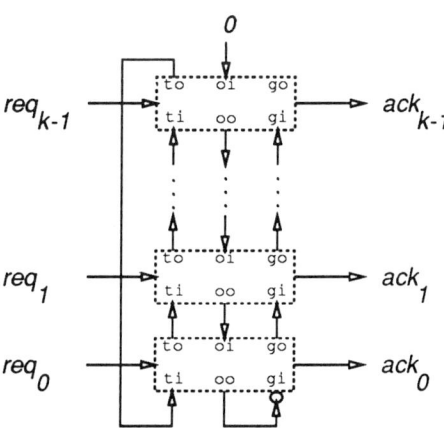

Figure 3.7 Configuration of the synchronous arbiter circuit

acknowledge outputs are asserted. The override signal propagates through the cells below, negating the grant input of cell 0, and thus preventing any other cells from acknowledging at the same time. The circuit is initialized so that all of the W registers are reset and exactly one T register is set.

The desired properties of the arbiter circuit are:

1. No two acknowledge outputs are asserted simultaneously
2. Every persistent request is eventually acknowledged
3. Acknowledge is not asserted without request

Expressed in CTL, they are:

1. $\bigwedge_{i \neq j} AG \neg(\text{ack}_i \wedge \text{ack}_j)$
2. $\bigwedge_i AGAF(\text{req}_i \Rightarrow \text{ack}_i)$
3. $\bigwedge_i AG(\text{ack}_i \Rightarrow \text{req}_i)$

Using the symbolic CTL model checking procedure, we can determine whether the design has these properties, for a given number of cells. Figure 3.8 plots the performance of the symbolic model checking procedure for this example in terms of several measures: the size of the transition relation in OBDD nodes, the total run time (on a Sun3, running an implementation in the C language), and the maximum number of OBDD nodes used at any given time.[2] We observe that as the number of cells in the circuit increases, the size of the transition relation increases linearly (in section 3.5, we will prove a theorem that shows why this is the case). The execution time is well fit by a quadratic curve. The number of reachable states, however, explodes exponentially (note the logarithmic scale on the reachable states axis).

To obtain polynomial performance for this example, it was necessary to add a wrinkle to the symbolic model checking algorithm. In the first experiment it was found that although most of the specification was checked quickly, the time required to check property 2 for cell 0 doubled each time a cell was added. The reason for this is rather remarkable. Consider a function called *Rotate*, which returns true for a pair of n bit binary numbers when one number can be obtained from the other by a rotation of j bits. There is no variable ordering which yields an efficient OBDD for this function for all j.[3] In fact, a very similar function occurs in computing the set of states satisfying the formula $AF(\text{req}_0 \Rightarrow \text{ack}_0)$, where the two binary numbers are given by the W and T registers respectively. Note that, for cell 0, request implies acknowledge exactly when no other cell has both W and T registers set. The T registers rotate once per clock cycle. Thus, $\text{req}_0 \Rightarrow \text{ack}_0$ is necessarily true j steps in the future exactly when there is no other cell i for which $W_i \wedge T_{i-j \bmod k}$. The OBDD representing this set of states grows exponentially in the number of cells.

This illustrates a fairly general phenomenon: circuits tend to be "well behaved" in the part of their state space which is reachable from the initial state, but not elsewhere. In the case of the synchronous arbiter, only states with one T register set are reachable. However, the symbolic model checking technique considers all states, including states with multiple tokens. A good solution to this problem in general is first to compute the set of reachable states, and then to restrict all of computations of the CTL model checking algorithm to those states. Since the reachable states are closed under the transition relation, this has no effect on the truth value obtained for formulas at the initial state. In particular, this solves the problem of the bus arbiter circuit, since in its reachable state space,

---

[2] The latter number should be regarded as being accurate only to within a factor of two, since the garbage collector in the implementation scavenges for unreferenced nodes only when the number of nodes doubles.

[3] This can be shown using the technique of [Bry91]. It is sufficient that for any variable order there is some rotation such that when the order is cut in half, information proportional to n must be passed from one half to the other.

Symbolic model checking 43

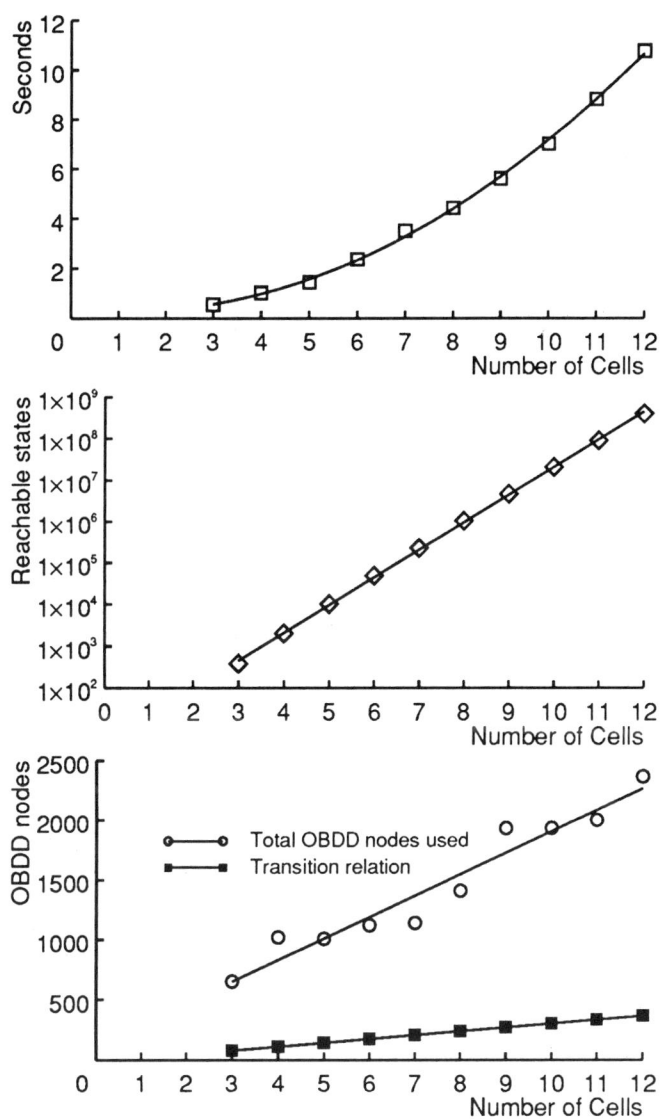

Figure 3.8 Performance − synchronous arbiter example.

the T registers cannot represent an arbitrary binary number.

The set of reachable states can be computed as the limit of a fixed point series. In particular, it is

$$\mu Q.\lambda y.(I(y) \vee \exists y'.(R(y', y) \wedge Q(y')))$$

where I is the set of initial states. The fixed point series starting with false gives us the sequence of sets Q_i reachable after i or fewer steps. In effect, this is a breadth-first search of the state space. In our symbolic representation, we have

$$\mu\mathbf{Q}.(\lambda\bar{v}'.(\mathbf{I} \vee \exists\bar{v}.(\mathbf{R} \wedge \mathbf{Q})))(\bar{v})$$

By computing the reachable states first and then using this set to restrict the CTL model checking algorithm, we obtain the polynomial run time results described above. This technique is also used for other experiments described in the sequel, unless otherwise noted.

3.4.2 Asynchronous state machines

In an asynchronous state machine, there is no global clock to which all state changes are synchronized. This makes designing correct asynchronous circuits considerably more challenging than designing correct synchronous circuits. We will consider two plausible models of asynchronous state machines. In the first, which we will call the *simultaneous model*, any or all state variables may change state in a given transition. Each state component makes an independent and non-deterministic choice regarding whether to change value or not. In the second model, which we will call the *interleaving model*, only one state component changes value in a given transition. The choice of which component changes value is non-deterministic.[4] In either model, we consider an asynchronous state machine composed of n gates. We will use state variable v_i to stand for the output of gate i, and f_i to represent the Boolean function computed by gate i.

In the simultaneous model, the transition relation can be represented by a formula in the form:

$$\mathbf{R} = \bigwedge_{1 \leq i \leq n} \mathbf{R}_i, \text{ where } \mathbf{R}_i = (v_i' \iff f_i) \vee (v_i' \iff v_i). \tag{3.12}$$

---

[4] A discussion of which state machine model is more suitable for circuit design is beyond the scope of this work. In general, conditions would have to be imposed on either model in order to make it implementable in a given design style. For discussion of asynchronous design techniques, see [MB59, Sei80b].

Symbolic model checking

For any transition and any state variable v_i, either the new value of v_i is determined by f_i, or it is the same as the old value. Note that this differs from the synchronous model (3.11) in which every state variable is reevaluated at every transition.

In the interleaving model, the transition relation can be represented by a formula in the form:

$$\bigvee_i R_i, \quad \text{where } \mathbf{R}_i = (v_i' \iff f_i) \wedge (\bigwedge_{j \neq i}(v_j' \iff v_j)) \qquad (3.13)$$

In any transition, *for some* state variable v_i, the new value of v_i is determined by f_i, and the remaining variables keep their old value. Note that in this case, the transition relation is represented by a *disjunction* of component relations rather than a conjunction.

In general, for models of parallel processes whose actions interleave arbitrarily, the transition relation is disjunctive. If this is the case, we can make an easy optimization in the symbolic model checking technique: we observe that the set of states reachable by one step of the system is the union of the sets of states reachable by one step of each individual component. This is reflected in the fact that existential quantification distributes over disjunction. Thus:

$$\begin{aligned} EX\mathbf{p} &= \exists \bar{v}'. ((\bigvee_i \mathbf{R}_i) \wedge \mathbf{p}') \\ &= \bigvee_i \exists \bar{v}'. (\mathbf{R}_i \wedge \mathbf{p}') \end{aligned}$$

Using this equality, we can avoid computing the transition relation of the system and instead use only the transition relations of the individual processes. This technique is called early quantification[5] – by rearranging the computations, we apply quantification before the logical disjunction operation. Heuristically, quantification tends to reduce OBDD size, since it reduces the number of variables. Hence, the size of the intermediate results is usually reduced (though the final result is the same).

Our example of an asynchronous state machine is the distributed mutual exclusion (DME) circuit of Alain Martin [Mar85]. It is a speed-independent circuit [Sei80b] and makes use of special two-way mutual exclusion circuits as components. Figure 3.9 is a diagram of a single cell of the distributed mutual-exclusion ring. The circuit works by passing a token around the ring, via the

---
[5] The *AndExists* algorithm of section 3.3, which combines conjunction and quantification in a bottom-up manner is also an example of early quantification.

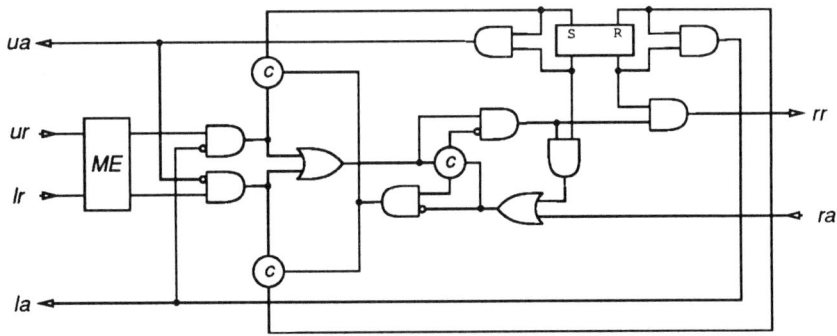

Figure 3.9 One cell of the DME circuit

request and acknowledge signals lr and la on the left and rr and ra on the right. A user of the DME gains exclusive access to the resource via the request and acknowledge signals ur and ua.

The specifications of the DME circuit are as follows:

1. No two users are acknowledged simultaneously.

2. An acknowledgment is not output without a request.

3. An acknowledgment is not removed while a request persists.

4. All requests are eventually acknowledged.

We will consider only the first specification, regarding mutual exclusion. The others are easily formulated in CTL, although the last requires the use of fairness constraints (see section 6.4) to guarantee that all gate delays are finite. The formalization of the mutual exclusion specification is

$$\bigwedge_{i \neq j} AG \neg (ua_i \wedge ua_j)$$

Now let's look at the performance of the symbolic model checking algorithm in checking this formula, for both a simultaneous and an interleaving model of the circuit. For the interleaving model, we use the early quantification technique. Figure 3.10 plots the relative performance for the simultaneous model (method 1) and the interleaving model (method 2). Part (a) shows the run time as a function of the number of DME cells, part (b) shows the total storage used (measured in OBDD nodes) and part (c) shows the number of nodes used to represent the transition relation. For the moment, disregard the curves for method 3. The experiment was run for up to 7 cells of the simultaneous model (limited by space) and up to 10 cells of the interleaving model (limited by time). Part (b) of the figure shows the substantial space advantage of the interleaving model, and from part (c), we can see that most of the difference is accounted for by the savings in representing the transition relation using early quantification. In both cases, the space used is linear in the number of cells. However, we note that the increase in run time appears to be cubic for the simultaneous model, but quartic for the interleaving model. It would seem that if enough storage were available to continue the curve for method 1, the two curves would meet in the neighborhood of 10 cells.

The different asymptotic performance for the simultaneous and interleaving models can be understood by looking at the OBDDs that occur in the fixed point iterations computing the reachable states. Figure 3.11 plots the size of the largest such OBDD for each method. We can see clearly that the size is increasing linearly for the simultaneous model, but quadratically for the interleaving model. This is a phenomenon which occurs generally when comparing simultaneous *vs.* interleaving models. It can be understood by considering a very simple system composed of n processes, each with states 0 and 1, and each alternating non-deterministically between these two states. If we start the system with all processes in state 0, what do we observe after k steps? In the simultaneous case, after one step, all possible states are reachable. In the interleaving case, however, after k steps, all global states with at most k 1's are reachable. This is a symmetric function. As Bryant noted [Bry86], all symmetric functions can be represented by a quadratic size OBDDs. The symmetry results from the fact that in an interleaving model, exactly one state component changes in a given transition, and the choice is arbitrary. In general, after k steps of such a model, the number of steps taken by each state component sums to k. Hence, in the set of states reachable after k steps, there is an induced correlation between the states of otherwise independent processes.

The simultaneous model appears to be inferior to the interleaving model from a symbolic model checking point of view, owing to the large amount of space required to represent the transition relation. Most of this, however, can be attributed to a phenomenon we observed in the previous example: systems tend to be well behaved only in their reachable state space. In the symbolic model checking technique, we represent the transition relation over the entire state space. Although representing only the reachable transitions might be more efficient, we seem to be caught in Catch 22: we need to represent the transition relation to compute the set of reachable states. We can avoid this problem by incrementally computing only as much of the transition relation as is necessary to compute the next iteration of the fixed point algorithm. The reachable state set is the least fixed point of $\tau = \lambda Y.(I \vee R(Y))$, where $R(Y)$ is the image of Y via R. By rearranging the fixed point computation slightly, we only need represent R correctly for those transitions (x, y), where x is on the "frontier" of the search:

$$\begin{aligned}\tau^{i+1}(\text{false}) &= \tau^i(\text{false}) \vee R(\tau^i(\text{false})) \\ &= \tau^i(\text{false}) \vee R(\tau^i(\text{false}) - \tau^{i-1}(\text{false}))\end{aligned}$$

At each iteration, we can reevaluate the formula R over the set of states $\tau^i(\text{false}) - \tau^{i-1}(\text{false})$. This can be done by restricting each subformula using either the logical *and* or using the *Restrict* operator of Coudert, Madre and Berthet (see section 6.5). This results in a sequence of approximations to the transition relation which are substantially more compact than the complete transition relation, although we must reevaluate R at each iteration, rather that evaluating it once at the beginning. We will call this method 3.

In part (a) of figure 3.10, we see that the time used by this method, while still cubic, is a substantial improvement over the previous method for the simultaneous model (method 1). More importantly, the space used is dramatically improved, allowing a model with a larger number of cells to be checked. The method overtakes the interleaving model in run time at about 8 cells, owing to its better asymptotic performance.

Figure 3.12 plots the number reachable states as a function of the number of cells (the numbers are indistinguishable for the two models). The number of reachable states grows exponentially in the number of cells, though not as rapidly as the total number of states, which is 2^{18n}. The key point is that for all three methods, the space and time necessary for the symbolic model checking method is polynomial in the number of cells. Thus, the state explosion problem has been avoided. The overall time complexity of $O(n^3)$ for the simultaneous model derives from three factors: a linear increase in the transition relation OBDD, a linear increase in the state set OBDDs obtained in the fixed point iterations, and a linear increase in the number of iterations. For the interleaving model, the quadratic increase in the state set OBDDs results in an overall $O(n^4)$ time complexity. On the other hand, the number of reachable states increases roughly a factor of ten with each added cell.

It is not immediately clear that either the interleaving or simultaneous model is preferable in general. Interleaving models seem to be better when the number

Symbolic model checking

of asynchronous processes is small, and simultaneous when the number is large. The cache consistency protocol of chapter 5 is an example of a large system with a fairly small number of complex asynchronous processes. This is an appropriate application of an interleaving model.

The polynomial performance of the symbolic model checking algorithm, in spite of the exponential increase in states, makes it possible to analyze fairly large instantiations of the two example circuits (the synchronous arbiter and the DME circuit). It should be possible to verify these and similar circuits for any reasonable fixed number of cells. This begs the question – how many cells do we need to analyze to be guaranteed that the design is correct for any number of cells? Intuitively, for sufficiently large n, a sequence of $n + 1$ cells should be equivalent in some sense to a sequence of n cells. But in what sense equivalent? This problem is dealt with in chapter 7, where we consider induction over processes.

3.5 GRAPH WIDTH AND OBDDS

In this section, we consider the asymptotic growth of OBDDs representing certain topological classes of circuits. This analysis explains some of the performance results of the previous section.

In 1989, Berman proved a bound on the OBDD size needed to represent circuits of bounded width. A circuit has bounded width if its elements can be arranged in a linear order such that any cut through the order crosses at most a bounded number of wires w, called the width of the circuit. There exists a variable ordering such that the OBDD size is bounded by $n2^w$, where n is the number of primary inputs of the circuit. This result applies only if the order is "topological", meaning essentially that the direction of all the wires follows the ordering. Here, this result is generalized, to show that if w_f bounds the number wires through any cut in the forward direction, and w_r bounds the number in the reverse direction, then the OBDD size is bounded by $n2^{w_f}2^{w_r}$. For the case where $w_r = 0$, this is the same as Berman's result. Using this result, we can linearly bound the OBDD representation for the transition relation of circuits like the arbiter and the DME ring, which have linear arrangements with a bounded number of wires through any crosss section.

Fujita states that tree circuits using only AND, OR and XOR gates have linearly bounded OBDD representations [FMK90]. Here, we show that a more

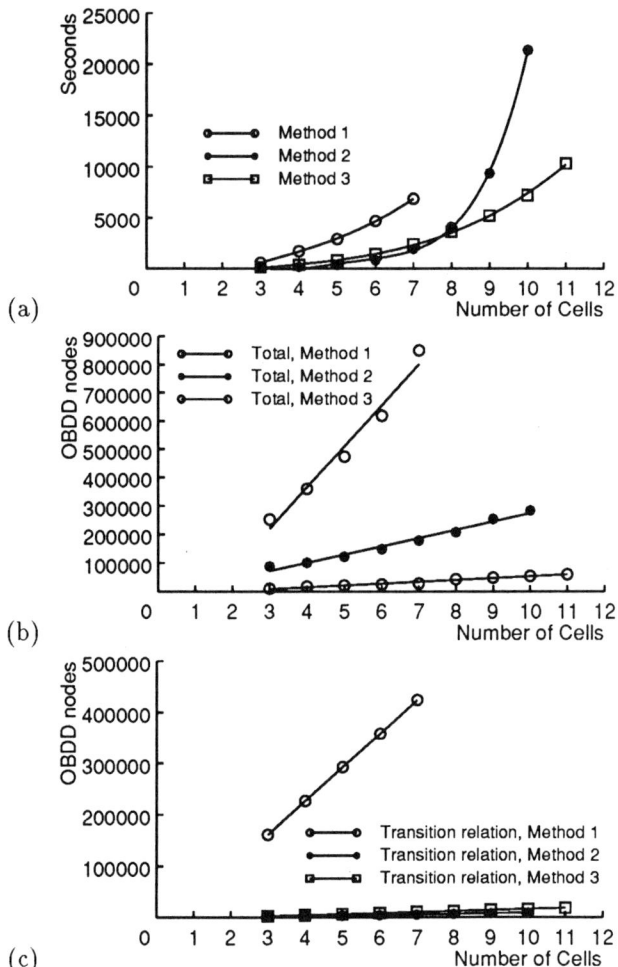

Figure 3.10 Performance for DME circuit example

Symbolic model checking 51

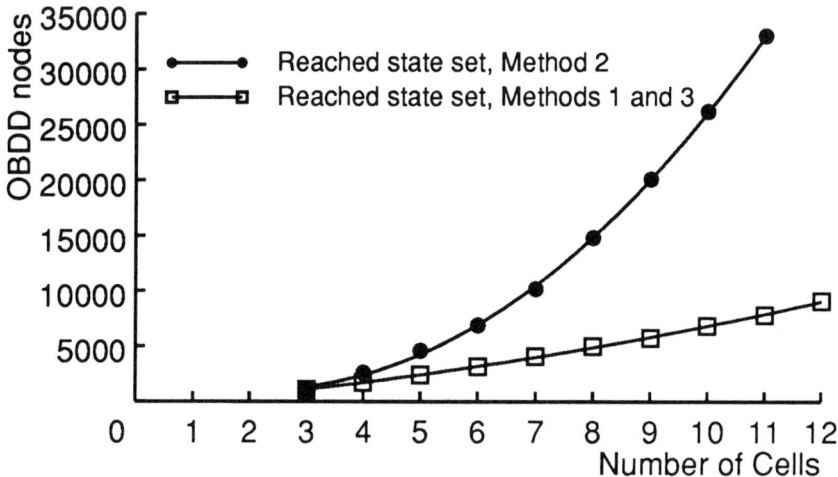

Figure 3.11 State set size for DME circuit example

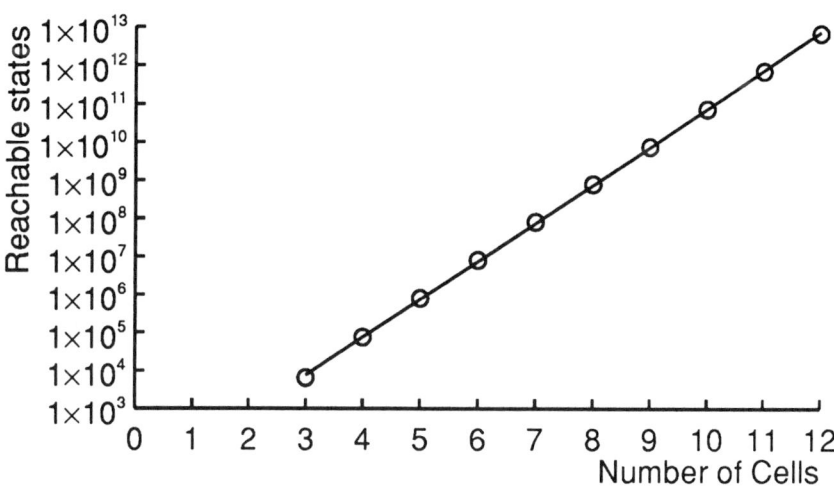

Figure 3.12 Reachable states for DME circuit example

general class of circuits with bounded "tree width" and arbitrary function elements have polynomially bounded OBDDs. The essence of the argument is to show that these circuits can be arranged in a linear order with a width that is logarithmic in the number of gates. This yields a bound on the OBDD size which is polynomial in the number of gates.

3.5.1 Bounded width circuits

Let $L = (G, <)$ be a linear order on the gates of a circuit. We classify the primary inputs and outputs of the circuit as special instances of gates in order to simplify the definitions, and assume that the primary output is at the top of the order. Given an order L, we will say that the *forward cross section* of the circuit at gate g is the set of wires connected to an output of some gate g_1 and an input of some gate g_2 such that $g_1 \leq g$ and $g < g_2$. The *reverse cross section* is the set of wires connected to an output of some gate g_1 and an input of some gate g_2 such that $g_2 \leq g$ and $g < g_1$. We assume that no wire is connected to the outputs of two distinct gates, so these two sets are disjoint. We also assume that there are no cycles in the circuit, to insure that the circuit computes a function. The order L is said to be *topological* when all of the reverse cross sections are empty.

The forward width of the circuit under order L, denoted w_f, is the maximum size of the forward cross section at any gate g. Similarly, the reverse width of the circuit under order L, denoted w_r is the maximum size of the reverse cross section at any gate g.

The cross section of an OBDD at level i is the set of nodes labeled with variable v_i. The width w_p of an OBDD p is the maximum size of any cross section of p. The size of an OBDD is the sum of the sizes of its cross sections. Thus, the OBDD size if bounded by $n \cdot w_p$, where n is the number of variables.

It is easily shown that the size of the cross section of an OBDD at level i is the number of distinct functions

$$f_{px}(v_i, \ldots, v_n) = f_p(x_1, \ldots, x_{i-1}, v_i, \ldots, v_n)$$

which depend on v_i, where $x = (x_1, \ldots, x_{i-1})$ is a Boolean vector and f_p is the function represented by p. This observation leads to the following theorem bounding the size of an OBDD in terms of the forward and reverse widths of the circuit it represents:

Symbolic model checking

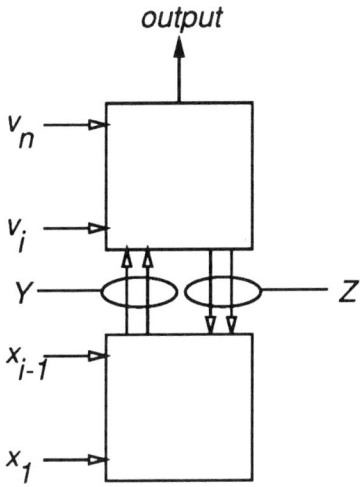

Figure 3.13 Proof of bounded width theorem

Theorem 4 *If a circuit computing function f has forward width w_f and reverse width w_r for some linear order L, then there is an OBDD p representing function f of size bounded by $n2^{w_f} 2^{w_r}$, where n is the number of inputs of the circuit.*

Proof. Associate the variables $v_1, \ldots, v_n$ of the OBDD with the inputs of the circuit, such that for all $i \leq j$, $v_i \leq v_j$. We can bound the size of the ith cross section of the resulting OBDD as follows. Let $x = (x_1, \ldots, x_{i-1})$ be a Boolean vector. Split the circuit in half by choosing any gate g such that $v_{i-1} \leq g < v_i$, letting Y be the forward cross section at g and Z the reverse cross section. This situation is depicted in figure 3.13. For any given value of x, Y is a function of Z, and this function determines $f_x(v_i, \ldots, v_n)$. The number of Boolean functions with $|Z|$ inputs and $|Y|$ outputs is $2^{|Y|2^{|Z|}}$ (to see this, count the number of entries in the truth table). This bounds the total number of distinct functions f_x, which in turn bounds the width of the OBDD representing f at level i. We know that $|Y| \leq w_f$ and $|Z| \leq w_r$. Thus, the overall OBDD size is bounded by $n \cdot 2^{w_f} 2^{w_r}$. □

This bound is linear in the number of inputs, exponential in the forward width and doubly exponential in the reverse width. The double exponential appears to be necessary. This can be shown using the "hidden weighted bit" function of [Bry91] as a counterexample. This circuit can be ordered in such a way that between any two inputs there is a cross section with $O(\log_2 n)$ wires in each direction, yet there is an exponential lower bound on its OBDD size. If we could bound the OBDD size with a single exponential in both the forward and reverse widths, the OBDD size would be $O(n2^{k \log_2 n}) = O(n^{k+1})$ where k is a constant.

The theorem is concerned with a single output of a combinational circuit, but it can also be applied to the transition relation of a sequential circuit. To do this, we simply transform the sequential circuit into a combinational circuit which computes the transition relation of the sequential circuit. This is done by adding a pair of inputs v_i and v_i' to represent the old and new values of each state component. Since the transition relation of the circuit is the conjunction of the transition relations of its components, we can do this while increasing the width of the circuit by only one wire in the forward direction as depicted in figure 3.14. Thus, for bounded width sequential circuits (even with wires in both directions), the size of the OBDD representing the transition relation is linear in $n_i + n_s$, where n_i is the number of inputs and n_s is the number of state components. The synchronous arbiter circuit and the DME circuit of the previous section provide experimental confirmation of this.

We have shown for a certain structural class of circuits that the representation of the transition relation is linearly bounded in the size of the circuit. We should note that in the symbolic model checking algorithm, we also use OBDDs to represent the set of states labeled with a given CTL formula. Unfortunately, we cannot expect to polynomially bound the size of the OBDDs representing these sets based purely on structural considerations. The simplest example of this is probably a circuit that inputs a binary number, stores one copy of it, then serially rotates the original by an arbitrary number of bits. This circuit has the simplest structure we might hope for that has any communication at all between the components, yet there is no variable order which yields a compact OBDD for the reachable state set of this circuit, since it implements the rotate function. The same argument would apply to a serial multiplier circuit. In general, if a circuit computes a function serially which cannot be represented by a compact OBDD, then we cannot expect the symbolic model checking algorithm using OBDDs to be efficient.

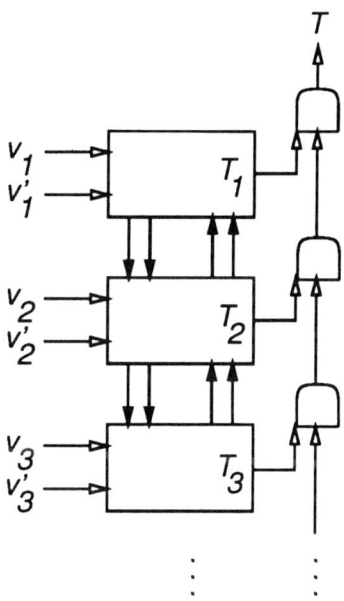

Figure 3.14 Computing a conjunctive transition relation

3.5.2 Bounded tree-width circuits

In the previous section, we considered the OBBD representation of circuits whose gates can be arranged in a sequence with a bounded number of wires in each cross section. Now we consider the slightly more general class of circuits which can be can be arranged in a tree with a bounded width property. This is not to say that the topology of the circuit must be a tree; rather, it must be possible to lay a spanning tree over the circuit in such a way that the width of the circuit across any arc of the spanning tree is bounded. This notion of bounded tree-width is defined as follows.

Let $T = (G, <)$ be a tree order over the gates of a circuit, where $g' < g$ iff g' is a descendant of g. Let b be the branching degree of T (ie., the maximum number of children of any gate). As before, the forward cross section at node g is the set of wires connecting an output of g_1 and an input of g_2 such that $g_1 \leq g$ and $g < g_2$. Similarly, the reverse cross section of T at node g is the set of wires connecting an output of g_1 and an input of g_2 such that $g_2 \leq g$ and $g < g_1$. The forward width of the tree w_f is the size of the largest forward cross section, while the reverse width w_r is the size of the largest reverse cross section.

For the moment, let us consider the case $w_r = 0$, and let the width w stand for the forward width:

Lemma 2 *For any topological tree order $T = (G, <)$, with width w and branching degree $b > 1$, there is a topological linear order $L = (G, <')$, with width $w' \leq w(b-1)log_2|G|$.*

Proof. By induction over $|G|$, the number of gates. The base case, $|G| = 1$, is trivial. Assume the theorem holds for all circuits of size less than $|G|$. Let g be the root of the tree, and let $G_1, \ldots, G_k$ be the subtrees of the root, where $k \leq b$, and $|G_1| \leq \cdots \leq |G_k|$. By inductive hypothesis, there exist linear orders $L_i = (G_i, <_i)$ of width $w_i \leq w(b-1)log_2|G_i|$, for all $1 \leq i \leq k$. Let $L = (G, <')$ be the extension of these orders such that $G_k <' \cdots <' G_1 <' g$, as depicted in figure 3.15. The width w' of L is bounded by $\max_{1 \leq i \leq k}(w_i + (k-i)w)$. Therefore, for some i,

$$w' \leq w_i + (k-i)w$$

In the case $k = i$, we have

$$w' \leq w_k \leq w(b-1)log_2|G_k| \leq w(b-1)log_2|G|$$

Symbolic model checking

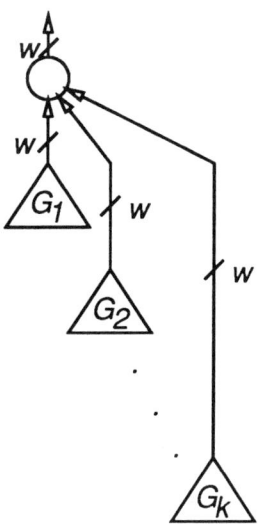

Figure 3.15 Arrangement of bounded width tree

Otherwise, $i < k$ and

$$\begin{aligned} w' &\leq w(k - i + (b-1)\log_2 |G_i|) \\ &\leq w(b-1)\log_2\left(2^{\frac{k-i}{b-1}}|G_i|\right) \end{aligned}$$

Here, we note that $|G_i| \leq (\Sigma_{i \leq j \leq k}|G_j|)/(k-i+1) \leq |G|/(k-i+1)$, so

$$w' \leq w(b-1)\log_2\left(\frac{2^{\frac{k-i}{b-1}}}{k-i+1}|G|\right)$$

We note that since $i \geq 1$ and $k \leq b$, $k - i \leq b - 1$. Therefore $2^{\frac{k-i}{b-1}} \leq 2$. Further, since $i < k$, $k - i + 1 \geq 2$. Thus $\frac{2^{\frac{k-i}{b-1}}}{k-i+1} \leq 1$. Therefore,

$$w' \leq w(b-1)\log_2 |G|$$

□

The theorem says that from any topological tree order of width w we can derive a linear order of width $w' \leq w(b-1)\log_2 |G|$. It follows by the previous theorem that the OBDD size is bounded by $n2^{w'} \leq n2^{w(b-1)\log_2 |G|} = n|G|^{w(b-1)}$, where

n is the number of primary inputs. This bound is polynomial in the size of the circuit for a fixed width and branching factor.

Now we turn to the question of tree orders that are not topological (*ie.*, bounded tree-width circuits with both forward and reverse wires). In this case, a logarithmic bound on the width of the linear order L is not sufficient, because the OBDD size can be doubly exponential in the number of *reverse* wires.

We can still obtain a polynomial bound in n, however, by converting a tree ordered circuit with reverse wires into a functionally equivalent tree ordered circuit with only forward wires:

Lemma 3 *If $T = (G, <)$ is a tree order over a circuit computing function f, with forward width w_f and reverse width w_r, then there is a circuit computing f with* topological *tree order $T' = (G', <')$ of forward width $w'_f \leq w_f 2^{w_r}$.*

Proof. Consider H, a subtree rooted at gate h, letting Y be the forward cross section at h, and Z the reverse cross section at h. Let $h_1, \ldots, h_k$ be the children of h, and let $Y_1, \ldots, Y_k$ and $Z_1, \ldots, Z_k$ be their respective forward and reverse cross sections. This situation is depicted in figure 3.16. Let the output functions computed by H be

$$Y = f(Z, Y_1, \ldots, Y_k)$$

and for $1 \leq i \leq k$, let

$$Z_i = r_i(Z, Y_1, \ldots, Y_k)$$
$$Y_i = f_i(Z_i)$$

We show by induction over $|H|$ that there exists a tree circuit H' of forward width $w'_f \leq w_f 2^{w_r}$ and reverse width $w'_r = 0$, computing the functions

$$f_x = f(x, Y_1, \ldots, Y_k), \text{ for } x \in \{0,1\}^{|Z|}$$

Note that f_x is simply row x in the truth table for Y. Since there are $2^{|Z|}$ possible values of x, and f_x has $|Y|$ components, the number of outputs of H' is $|Y|2^{|Z|}$.

By inductive hypothesis, there exist circuits H'_i for $1 \leq i \leq k$, satisfying the width bound and computing the functions

$$f_{ix} = f_i(x), \text{ for } x \in \{0,1\}^{|Z_i|}$$

Symbolic model checking

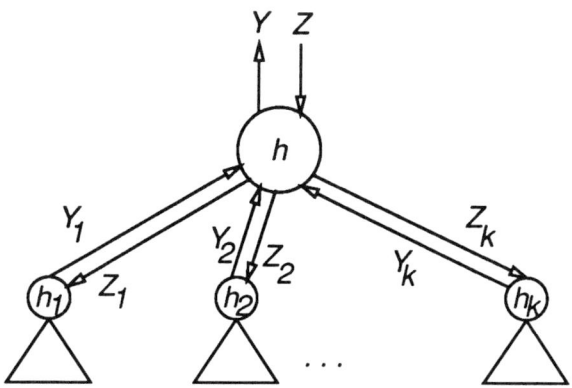

Figure 3.16 A non-topological tree order

Now, let h' be a gate computing f_x according to the following system of equations:

$$f_x = f(x, f_{1x_1}, \ldots, f_{kx_k})$$
$$x_i = r_i(x, f_{1x_1}, \ldots, f_{kx_k}), \text{ for } 1 \leq i \leq k$$

Let H' be the tree ordered circuit obtained by taking h' as the root, and $H'_1, \ldots, H'_k$ as the children of the root. The reverse width at the root is 0, since f_{ix} does not depend on Z, and the forward width at the root is $|Y|2^{|Z|}$. Hence, using the inductive hypothesis, $w'_r = 0$ and $w'_f \leq w_f 2^{w_r}$. If h is the root node of G, then H' computes the same function as G. □

This gives us the following theorem, bounding the OBDD size for tree ordered circuits with both forward and reverse wires:

Theorem 5 *If a circuit G computing function f has forward width w_f and reverse width w_r for some tree order T of branching degree $b > 1$, then there is an OBDD representing function f of size bounded by $n|G|^{w_f 2^{w_r}(b-1)}$, where n is the number of primary inputs of the circuit.*

Proof. According to lemma 3, for any tree ordered circuit of forward width w_f and reverse width w_r, we can construct a topological tree ordered circuit of width $w \leq w_f 2^{w_r}$, which computes the same function. By lemma 2, this circuit has a topological linear order L of width at most $w' \leq w(b-1)log_2|G|$. By theorem 4, there is an OBDD for the circuit of size bounded by

$$\begin{aligned} n2^{w'} &\leq n2^{w(b-1)\log_2|G|} \\ &\leq n2^{w_f 2^{w_r}(b-1)\log_2|G|} \\ &= n|G|^{w_f 2^{w_r}(b-1)} \end{aligned}$$

□

Hence, in the case of bounded tree width circuits (of a fixed branching degree), we also find that the OBDD size can be bounded polynomially in the size of the circuit. In this case, the exponent of n is related to both the width and the branching factor. Clearly, for this bound to be of any practical interest, w_f must be small, and w_r must be very small. Nonetheless, the theorem demonstrates a more general topological class of circuits with asymptotically compact OBDDs than was previously known.

4
THE SMV SYSTEM

In order to apply symbolic model checking to real problems, we need expressive languages that we can use to describe our model at a suitably high level (*eg.*, a gate level schematic is probably not a high enough level). For our purposes, this means the language must provide operations on suitable high level types (such as symbolic enumerated types), and must allow us to conveniently describe non-deterministic choices, so that we can describe high level protocols without being concerned with implementation details. The language must have a precise mathematical semantics that defines the translation from a program in the langauge to a form suitable for symbolic model checking (*ie.*, a Boolean formula representing the transition relation). For reasons that will be clarified in chapter 7, the semantics should be syntax-directed (defining the meaning of a language construct in terms of the meanings of its syntactic parts), so that we may infer properties of a program from properties of its parts. Ideally, it should also be possible to translate at least a useful subset of the language (not including, for example, non-deterministic choice) into hardware descriptions suitable for synthesis into real hardware. Constructs should be provided to allow us to use most common styles of system design.

Clearly, with all of these requirements, certain compromises will have of be made in the design of such a language, and the associated software for verification. The SMV system is one such compromise. It is a tool for checking finite state systems against specifications in the temporal logic CTL. The input language of SMV is designed to allow the description of finite state systems that range from completely synchronous to completely asynchronous, and from the detailed to the abstract. One can readily specify a system as a synchronous Mealy machine, or as an asynchronous network of abstract, nondeterministic processes. The language provides for modular hierarchical descriptions, and for the definition of reusable components. Since it is intended to describe finite state machines, the only basic data types provided by the language are bounded integer subranges and symbolic enumerated types. Static, structured data types can also be constructed from thes basic components. The logic CTL allows a rich class of temporal properties, including safety, liveness, fairness and deadlock freedom, to be specified in a concise syntax. SMV uses the OBDD-based symbolic model checking algorithm to efficiently determine whether specifications expressed in CTL are satisfied.

It is a truism that a language should be designed to enforce a style of description that avoids certain common kinds of errors. A language allowing a system's transition behavior to be described by any Boolean formula would be quite flexible, but also open to a great deal of abuse that would render the verification results vacuous. For example, such a formula might contain a logical contradiction. In this case, all specifications with universal path quantifiers would be vacuously true. The user would then be satisfied that a correct design had been obtained, only to find later that the design was unimplementable due to its inherent logical inconsistency. A wiser course would be to design a language to be as flexible as possible, while still providing some guaranteee that programs can actually be implemented. The existence of at least one implementation means that verification results are not purely vacuous.

Our approach to this problem will be to base the language on a notion of synchronous dataflow, which is closely related to the synchronous style of digital system design, but with higher level operations and non-deterministic choice. This makes the language similar in style to the LUSTRE language [CHPP87], though there are some distinguishing features. A program in this style is made up of a collection of parallel assignment statements, which may or may not involve a unit of delay. Such a program is implementable if it can be ordered by dataflow without zero-delay dependency cycles. This is somewhat like saying that we can build a circuit implementing the program from zero delay gates and unit delay latches. Asynchronous design styles can be simulated in this style by introducing processes that have arbitrary delay rather than unit delay.

This chapter describes the syntax and semantics of the SMV language in terms appropriate to symbolic model checking. This language is used in the following chapter to describe the cache consistency protocol of a multiprocessor so that properties of the protocol can be checked using the symbolic model checking technique.

4.1 AN INFORMAL INTRODUCTION

Before delving into the syntax and semantics of the language, let us first consider a few simple examples that illustrate the basic concepts. Consider, for example, the following short program in the language.

```
MODULE main
VAR
  request : boolean;
  state : {ready,busy};
ASSIGN
  init(state) := ready;
  next(state) := case
                    state = ready & request : busy;
                    1 : {ready,busy};
                 esac;
SPEC
  AG(request -> AF state = busy)
```

The program describes both the model and the specification. The model is a Kripke model, whose state is defined by a collection of state variables, which may be of Boolean, integer subrange, or enumerated type. The variable **request** is declared to be a Boolean in the above program, while the variable **state** can take on the symbolic values **ready** or **busy**. The value of an enumerated type variable is encoded by the compiler using a collection of Boolean variables, so that the transition relation may be represented by an OBDD. The exact manner in which this is done, however, is not visible to the user.

The transition behavior of the program, and its initial state (or states), are determined by a collection of parallel assignments. We can view these as simultaneous equations which, when solved, tell us what state transitions the program can make. Assignments are introduced by the keyword ASSIGN. In the above program, the initial value of the variable **state** is set to **ready**. At any given time, the value of **state** one time unit in the future is given by the expression

```
case
  state = ready & request : busy;
  1 : {ready,busy};
esac;
```

The value of a **case** expression is determined by the first expression on the right hand side of a (:) such that the condition on the left hand side is true. Thus, if **state = ready & request** is true, then the result of the expression is **busy**, otherwise, it is the set {**ready,busy**}. When a set is assigned to a variable, the result is a non-deterministic choice among the values in the set. Thus, if the value of **status** is not **ready**, or **request** is false (in the current state), the value of **state** in the next state can be either **ready** or **busy**. Non-deterministic choices are useful for describing systems which are not yet fully implemented (*ie.*, where some design choices are left to the implementor), or abstract models of complex protocols, where the value of some state variables cannot be completely determined.

Notice that the variable **request** is not assigned in this program. This leaves the SMV system free to choose any value for this variable, giving it the characteristics of an unconstrained input to the system.

The specification of the system appears as a formula in CTL under the keyword **SPEC**. The SMV model checker verifies that all possible initial states satisfy the specification. In this case, the specification is that invariantly if **request** is true, then inevitably the value of **state** is **busy**.

The following program illustrates the definition of reusable modules and expressions. It is a model of a 3 bit binary counter circuit. Notice that the module name "**main**" has special meaning in SMV, in the same way that it does in the C programming language. The order of module definitions in the input file is inconsequential.

```
MODULE main
VAR
   bit0 : counter_cell(1);
   bit1 : counter_cell(bit0.carry_out);
   bit2 : counter_cell(bit1.carry_out);
SPEC
   AG AF bit2.carry_out

MODULE counter_cell(carry_in)
VAR
   value : boolean;
ASSIGN
   init(value) := 0;
   next(value) := value + carry_in mod 2;
DEFINE
   carry_out := value & carry_in;
```

The SMV system

In this example, we see that a variable can also be an instance of a user defined module. The module in this case is `counter_cell`, which is instantiated three times, with the names `bit0`, `bit1` and `bit2`. The counter cell module has one formal parameter `carry_in`. In the instance `bit0`, this formal parameter is given the actual value 1. In the instance `bit1`, `carryin` is given the value of the expression `bit0.carry_out`. This expression is evaluated in the context of the main module. However, an expression of the form $a.b$ denotes component b of module a, just as if the module a were a data structure in a standard programming language. Hence, the `carry_in` of module `bit1` is the `carry_out` of module `bit0`. The keyword **DEFINE** is used to assign the expression `value & carry_in` to the symbol `carry_out`. Definitions of this type are useful for describing Mealy machines. They are analogous to macro definitions, but notice that a symbol can be referenced before it is defined.

The effect of the DEFINE statement could have been obtained by declaring a variable and assigning its value, as follows:

```
VAR
   carry_out : boolean;
ASSIGN
   carry_out := value & carry_in;
```

Notice that in this case, the *current* value of the variable is assigned, rather than the next value. Defined symbols are sometimes preferable to variables, however, since they don't require introducing a new variable into the OBDD representation of the system. The weakness of defined symbols is that they cannot be given values non-deterministically. Another difference between defined symbols and variables is that while variables are statically typed, definitions are not. This may be an advantage or a disadvantage, depending on your point of view.

In a parallel-assignment language, the question arises: "What happens if a given variable is assigned twice in parallel?" More seriously: "What happens in the case of an absurdity, like `a := a + 1;` (as opposed to the sensible `next(a) := a + 1;`)?" In the case of SMV, the compiler detects both multiple assignments and circular dependencies, and treats these as errors, even in the case where the corresponding system of equations has a unique solution. Another way of putting this is that there must be a total order in which the assignments can be executed which respects all of the data dependencies. The same logic applies to defined symbols. As a result, all legal SMV programs are realizable as sequential circuits.

By default, all of the assignment statements in an SMV program are executed in parallel and simultaneously. It is possible, however, to define a collection of parallel processes, which run at arbitrary time intervals. This is useful for describing communication protocols, asynchronous circuits, or other systems whose actions are not synchronized to a global clock. This technique is illustrated by the following program, which represents a ring of three inverting gates.

```
MODULE main
VAR
   gate1 : process inverter(gate3.output);
   gate2 : process inverter(gate1.output);
   gate3 : process inverter(gate2.output);
SPEC
   (AG AF gate1.out) & (AG AF !gate1.out)

MODULE inverter(input)
VAR
   output : boolean;
ASSIGN
   init(output) := 0;
   next(output) := !input;
```

A *process* is an instance of a module which is introduced by the keyword **process**. At any given time, the process is either running or not running. An assignment to the **next** value of a variable only applies if the process is running. If not the value of the variable remains the same in the next time step. The interval between times that a process runs is arbitrary (*ie.*, non-deterministic). We use a **process** to model a gate in an asynchronous circuit, because we don't want to make any assumptions about how much time is required for the output of the gate to change state. In addition, at most one of the three processes in the above program is allowed to run at any given time. This gives us an interleaving model of execution, as in the example of section 3.4.2. The SMV model checker can take advantage of this by structuring the Binary Decision Diagrams to give the effect of a disjunctive transition relation. This can yield a substantial savings in space in representing the transition relation.

The specification of this program states that the output of **gate1** must oscillate (*ie.*, that its value is infinitely often zero, and infinitely often one). In fact, this specification is false, since the system is not forced to execute a process infinitely often. Hence, the output of a given gate may remain constant, regardless of changes of its input.

The SMV system

In order to force a given process to execute infinitely often, we can use a *fairness constraint*. A fairness constraint restricts the attention of the model checker to those execution paths along which a given CTL formula is true infinitely often. Each process has a special variable called **running** which is true if and only if that process is currently executing. By adding the declaration

```
FAIR
   running
```

to the module **inverter**, we can effectively force every instance of **inverter** to execute infinitely often, thus making the specification true.

The alternative to using processes to model an asynchronous circuit would be to have all gates execute simultaneously, but allow each gate the non-deterministic choice of whether or not to change its output. This gives us the effect of a *simultaneous model*, using a conjunctive transition relation. Such a model of the inverter ring would look like the following:

```
MODULE main
VAR
   gate1 : inverter(gate3.output);
   gate2 : inverter(gate2.output);
   gate3 : inverter(gate1.output);
SPEC
   (AG AF gate1.out) & (AG AF !gate1.out)

MODULE inverter(input)
VAR
   output : boolean;
ASSIGN
   init(output) := 0;
   next(output) := !input union output;
```

The union operator allows us to express a nondeterministic choice between two expressions. Thus, the next output of each gate can be either its current output, or the negation of its current input – each gate can choose non-deterministically whether to delay or not. As a result, the number of possible transitions from a given state can be as high as 2^n, where n is the number of gates. This sometimes (but not always) makes it more expensive to represent the transition relation. The relative advantages of interleaving and simultaneous models of asynchronous systems are discussed in section 3.4.2.

The parallel assignment mechanism has the advantage that programs using are guaranteed to be implementable. However, should the need arise, it is possible in SMV to specify the transition relation directly using a Boolean formula in terms of the current and next values of the state, variables. Similarly, it is possible to give an initial condition in terms of the current state variables. These two functions are accomplished by the TRANS and INIT statements respectively. As an example, here is a description of the three inverter ring using only TRANS and INIT:

```
MODULE main
VAR
   gate1 : inverter(gate3.output);
   gate2 : inverter(gate1.output);
   gate3 : inverter(gate2.output);
SPEC
   (AG AF gate1.out) & (AG AF !gate1.out)

MODULE inverter(input)
VAR
   output : boolean;
INIT
   output = 0
TRANS
   next(output) = !input | next(output) = output
```

According to the TRANS declaration, for each inverter, the next value of the output is equal either to the negation of the input, or to the current value of the output. Thus, in effect, each gate can choose non-deterministically whether or not to delay. The use of TRANS and INIT is not recommended, since logical absurdities in these declarations can lead to unimplementable descriptions. For example, one could declare the logical constant 0 (false) to represent the transition relation, resulting in a system with no transitions at all. However, the flexibility of these mechanisms may be useful for those writing translators from other languages to SMV.

To summarize, the SMV language is designed to be flexible in terms of the styles of models it can describe. It is possible to fairly concisely describe synchronous or asynchronous systems, to describe detailed deterministic models or abstract nondeterministic models, and to exploit the modular structure of a system to make the description more concise. It is also possible to write logical absurdities if one desires to, and also sometimes if one does not desire to, using the TRANS and INIT declarations. By using only the parallel assignment mechanism, however, this problem can be avoided. The language is designed to exploit the capabilities of the symbolic model checking technique. As a result the available data types are all static and finite. No attempt has been made to support a particular model of communication between concurrent processes (*eg.*, synchronous or asynchronous message passing). In addition, there is no explicit support for some features of communicating process models such as sequential execution. Since the full generality of the symbolic model checking technique is available through the SMV language, it is possible that translators from various languages, process models, and intermediate formats could be created. In particular, existing silicon compilers could be used to translate high level languages with rich feature sets into a low level form (such as a Mealy machine) that could be readily expressed in the SMV language.

4.2 THE INPUT LANGUAGE

This section describes the various constructs of the SMV input language, and their syntax.

4.2.1 Lexical conventions

An atom in the syntax described below may be any sequence of characters in the set {A-Z,a-z,0-9,_,-}, beginning with an alphabetic character. All characters in a name are significant, and case is significant. Whitespace characters are space, tab and newline. Any string starting with two dashes ("--") and ending with a newline is a comment. A number is any sequence of digits. Any other tokens recognized by the parser are enclosed in quotes in the syntax expressions below.

4.2.2 Expressions

Expressions are constructed from variables, constants, and a collection of operators, including Boolean connectives, integer arithmetic operators, and case expressions. The syntax of expressions is as follows.

```
expr ::
        atom                    ;; a symbolic constant
        | number                ;; a numeric constant
        | id                    ;; a variable identifier
```

```
| "!" expr                ;; logical not
| expr1 "&" expr2         ;; logical and
| expr1 "|" expr2         ;; logical or
| expr1 "->" expr2        ;; logical implication
| expr1 "<->" expr2       ;; logical equivalence
| expr1 "=" expr2         ;; equality
| expr1 "<" expr2         ;; less than
| expr1 ">" expr2         ;; greater than
| expr1 "<=" expr2        ;; less that or equal
| expr1 ">=" expr2        ;; greater than or equal
| expr1 "+" expr2         ;; integer addition
| expr1 "-" expr2         ;; integer subtraction
| expr1 "*" expr2         ;; integer multiplication
| expr1 "/" expr2         ;; integer division
| expr1 "mod" expr2       ;; integer remainder
| "next" "(" id ")"       ;; next value
| set_expr                ;; a set expression
| case_expr               ;; a case expression
```

An **id**, or identifier, is a symbol or expression which identifies an object, such as a variable or defined symbol. Since an **id** can be an atom, there is a possible ambiguity if a variable or defined symbol has the same name as a symbolic constant. Such an ambiguity is flagged by the compiler as an error. The expression **next(x)** refers to the value of identifier **x** in the next state (see section 4.2.3). The order of parsing precedence from high to low is

```
*,/
+,-
mod
=,<,>,<=,>=
!
&
|
->,<->
```

Operators of equal precedence associate to the left. Parentheses may be used to group expressions.

A **case** expression has the syntax

```
case_expr ::
        "case"
```

The SMV system

```
        expr_a1 ":" expr_b1 ";"
        expr_a2 ":" expr_b2 ";"
        ...
"esac"
```

A case expression returns the value of the first expression on the right hand side, such that the corresponding condition on the left hand side is true. Thus, if **expr_a1** is true, then the result is **expr_b1**. Otherwise, if **expr_a2** is true, then the result is **expr_b2**, *etc.* If none of the expressions on the left hand side is true, the result of the case expression is the numeric value 1. It is an error for any expression on the left hand side to return a value other than the truth values 0 or 1.

A set expression has the syntax

```
set_expr ::
        "{" val1 "," val2 "," ... "}"
        | expr1 "in" expr2       ;; set inclusion predicate
        | expr1 "union" expr2    ;; set union
```

A set can be defined by enumerating its elements inside curly braces. The elements of the set can be numbers or symbolic constants. The inclusion operator tests a value for membership in a set. The union operator takes the union of two sets. If either argument is a number or symbolic value instead of a set, it is coerced to a singleton set.

4.2.3 Declarations

The VAR *declaration*

A state of the model is an assignment of values to a set of state variables. These variables (and also instances of modules) are declared by the notation

```
decl :: "VAR"
        atom1 ":" type1 ";"
        atom2 ":" type2 ";"
        ...
```

The type associated with a variable declaration can be either Boolean, an enumerated type, or a user defined module. A type specifier has the syntax

```
type :: boolean
        | "{" val1 "," val2 "," ... "}"
        | atom [ "(" expr1 "," expr2 "," ... ")" ]
        | "process" atom [ "(" expr1 "," expr2 "," ... ")" ]

val  :: atom | number
```

A variable of type **boolean** can take on the numerical values 0 and 1 (representing false and true, respectively). In the case of a list of values enclosed in set brackets (where atoms are taken to be symbolic constants), the variable is can take any of these values. Finally, an **atom** optionally followed by a list of expressions in parentheses indicates an instance of module **atom** (cf. section 4.2.4). The keyword **process** causes the module to be instantiated as an asynchronous process (cf. section 4.2.6).

The ASSIGN *declaration*

An assignment declaration has the form

```
decl :: "ASSIGN"
            dest1 ":=" expr1 ";"
            dest2 ":=" expr2 ";"
            ...

dest :: atom
        | "init" "(" atom ")"
        | "next" "(" atom ")"
```

On the left hand side of the assignment, **atom** denotes the current value of a variable, **init(atom)** denotes its initial value, and **next(atom)** denotes its value in the next state. If the expression on the right hand side evaluates to an integer or symbolic constant, the assignment simply means that the left hand side is equal to the right hand side. On the other hand, if the expression evaluates to a set, then the assignment means that the left hand side is contained in that set. It is an error if the value of the expression is not contained in the range of the variable on the left hand side.

In order for a program to be implementable, there must be some order in which the assignments can be executed such that no variable is assigned after its value is referenced. This is not the case if there is a circular dependency among the assignments in any given process. Hence, such a condition is an error. In addition, it is an error for a variable (either current or next value) to be assigned more than once, or for the initial value of a variable to be assigned more than once (although see exception in section 4.2.6).

The TRANS *declaration*

The transition behavior of the model is a defined by a Boolean formula on variables representing the current and next state of the program variables. This formula can be specified directly using the TRANS declaration. The syntax is

```
decl :: "TRANS" expr
```

The current value of a variable x is represented by just x, while the next value is represented by next(x). It is an error for the expression to yield any value other than 0 or 1. If there is more than one TRANS declaration, the transition relation is the conjunction of all of TRANS declarations.

The INIT *declaration*

The initial condition of the model is determined by a Boolean expression under the INIT keyword. The syntax of a INIT declaration is

```
decl :: "INIT" expr
```

It is an error for the expression to contain the next() operator, or to yield any value other than 0 or 1. If there is more than one INIT declaration, the initial condition is the conjunction of all of the INIT declarations.

The SPEC *declaration*

The system specification is given as a formula in the temporal logic CTL, introduced by the keyword SPEC. The syntax of this declaration is

```
decl :: "SPEC" ctlform
```

A CTL formula has the syntax

```
ctlform ::
    expr                          ;; a Boolean expression
    | "!" ctlform                 ;; logical not
    | ctlform1 "&" ctlform2       ;; logical and
    | ctlform1 "|" ctlform2       ;; logical or
    | ctlform1 "->" ctlform2      ;; logical implies
    | ctlform1 "<->" ctlform2     ;; logical equivalence
    | "E" pathform                ;; existential path quantifier
    | "A" pathform                ;; universal path quantifier
```

The syntax of a path formula is

```
pathform ::
    "X" ctlform                   ;; next time
    "F" ctlform                   ;; eventually
    "G" ctlform                   ;; globally
    ctlform1 "U" ctlform2         ;; until
```

The order of precedence of operators is (from high to low)

```
E,A,X,F,G,U
!
&
|
->,<->
```

Operators of equal precedence associate to the left. Parentheses may be used to group expressions. It is an error for an expression in a CTL formula to contain a **next()** operator or to return a value other than 0 or 1. If there is more than one **SPEC** declaration, the specification is the conjunction of all of the **SPEC** declarations.

The FAIR *declaration*

A fairness constraint is a CTL formula which is assumed to be true infinitely often in all fair execution paths. When evaluating specifications, the model checker considers path quantifiers to apply only to fair paths. Fairness constraints are declared using the following syntax:

```
decl :: "FAIR" ctlform
```

A path is considered fair if and only if all fairness constraints declared in this manner are true infinitely often along that path.

The DEFINE declaration

In order to make descriptions more concise, a symbol can be associated with a commonly used expression. The syntax for this declaration is

```
decl :: "DEFINE"
         atom1 ":=" expr1 ";"
         atom2 ":=" expr2 ";"
         ...
```

When every an identifier referring to the symbol on the left hand side occurs in an expression, it is replaced by the *value* of the expression on the right hand side (not the expression itself). Forward references to defined symbols are allowed, but circular definitions are not allowed, and result in an error.

4.2.4 Modules

A module is an encapsulated collection of declarations. Once defined, a module can be reused as many times as necessary. Modules can also be parameterized, so that each instance of a module can refer to different data values. A module can contain instances of other modules, allowing a structural hierarchy to be built. The syntax of a module is as follows.

```
module ::
         "MODULE" atom [ "(" atom1 "," atom2 "," ... ")" ]
                  decl1
                  decl2
                  ...
```

The atom immediately following the keyword MODULE is the name associated with the module. The optional list of atoms in parentheses are the formal parameters of the module. Whenever these parameters occur in expressions within the module, they are replaced by the actual parameters which are supplied when the module is instantiated.

An *instance* of a module is created using the VAR declaration (cf. section 4.2.3).

This declaration supplies a name for the instance, and also a list of actual parameters, which are assigned to the formal parameters in the module definition. An actual parameter can be any legal expression. It is an error if the number of actual parameters is different from the number of formal parameters. The semantics of module instantiation is similar to call-by-reference. For example, consider the following program fragment:

```
...
VAR
   a : boolean;
   b : foo(a);
...
MODULE foo(x)
ASSIGN
   x := 1;
```

The variable a is assigned the value 1. Now consider this program fragment:

```
...
DEFINE
   a := 0;
VAR
   b : bar(a);
...
MODULE bar(x)
DEFINE
   a := 1;
   y := x;
```

In this program, the value assigned to y is 0. Using a call-by-name (macro expansion) mechanism, the value of y would be 1, since a would be substituted as an expression for x.

Forward references to module names are allowed, but circular references are not, and result in an error.

4.2.5 Identifiers

An **id**, or identifier, is an expression which references an object. Objects are instances of modules, variables, and defined symbols. The syntax of an identifier is as follows.

The SMV system

```
id ::
        atom
      | id "." atom
```

The dot is used to form identifiers that correspond to the location of an object in the modular hierarchy of the program. For example, if an instance of a module is created with name a, then any occurrence of identifier x in the module (except as an actual parameter) becomes $a.x$ in the instantiation. The actual parameters are immune from this prefixing to give the effect of call-by-reference described above. As an example of using the dot operator in a program, consider the following fragment:

```
...
VAR
    a : foo(b);
    b : bar(a);
...
MODULE foo(x)
DEFINE
    c := x.p | x.q;

MODULE bar(x)
VAR
    p : boolean;
    q : boolean;
```

Here, we have passed the identifier of an instantce b as a parameter to instance a of module **foo**. This parameter is used as a prefix so that a can access components of b. This is rather like treating b as a variable of a structured type **bar**. This technique is useful in reducing the number of parameters when instances share a complex data object.

4.2.6 Processes

Processes are used to model interleaving concurrency. An instance of a module is identified as a process using the keyword **process** (cf. section 4.2.3). Each process has a special Boolean variable **running** which is true whenever the process is running. A process may run only when its parent is running. In addition, no two processes with the same parent may be running simultaneously. Apart from these restrictions, the value of **running** is non-deterministic.

An assignment of the form `ASSIGN next(v) := expr;` in a process applies only when **running** is true. When running is false, `next(v) = v`. There is an exception to this rule, however, which makes it possible for more than one process to modify a variable. An assignment of the form

```
ASSIGN if expr1 then   next(v) := expr1;
```

applies only when `expr1` is true. Otherwise the value of `next(v)` is undetermined. A variable may be conditionally assigned in this way in more than one place, provided the conditions are mutually exclusive.

4.2.7 Programs

The syntax of an SMV program is

```
program ::
       module1
       module2
       ...
```

There must be one module with the name **main** and no formal parameters. The module **main** is instantiated by the compiler.

4.3 FORMAL SEMANTICS

In this section we assign a formal semantics to SMV programs. Since we want to apply the symbolic model checking technique to SMV programs, it is most natural to describe the semantics of in terms of a translation from a program to a Boolean formula representing its transition relation. This has the added advantage of providing some insight into the function of the compiler in the SMV verification system.

It will be convenient to introduce some notations. First of all, λ notation will be useful for defining the semantics of modules. If x is a variable, and f a formula, then $\lambda x.f$ is a functional. It stands, in effect, for a function of one parameter, which is to be substituted for x in f. If g is another formula, then

$\lambda x.f(g)$ is equivalent to the term obtained by substituting g for x in f. The effect of λ can be summarized by the following axioms:

$$\lambda x.f(g) = f \quad \text{where } x \text{ does not occur in } f \quad (4.1)$$
$$\lambda x.x(f) = f \quad (4.2)$$
$$\lambda x.(\neg f)(g) = \neg \lambda x.f(g) \quad (4.3)$$
$$\lambda x.(f \vee g)(h) = (\lambda x.f(h)) \vee (\lambda x.g(h)) \quad (4.4)$$

These axioms can be used to eliminate λ's from from formulas by driving them inward until they meet variables.

Additionally, to handle non-Boolean values, we introduce the notation $a \rightarrow (b, c)$, which means roughly if a then b, else c, where b and c may be integers or symbolic constants.

Our semantics for SMV programs is simply a function that maps a given SMV program fragment f onto formula $[\![f]\!]$ which we call its *denotation*. The semantics is *syntax-directed*, meaning that the denotation of any program construct is defined as a function of the denotations of its syntactic components. In fact, an SMV program has different denotations for different purposes. For example, from a single program, we derive formulas for the transition relation, the initial condition, the fairness constraints and the specification. Where necessary, we will make this distinction by attaching a subscript, so that $[\![f]\!]_R$ represents the transition relation of f, $[\![f]\!]_I$ the initial condition, $[\![f]\!]_F$ the fairness constraints and $[\![f]\!]_S$ the specifications.

4.3.1 Semantics of variable declaration

The first syntactic category we will consider in our semantics is variable declarations. If n is a name, and l is a list of decl, then

$$[\![\texttt{VAR } n : \texttt{boolean}; \ l]\!] = [\![l]\!] \quad (4.5)$$

In other words, declaring a variable to be of type boolean is semantically neutral (although it does provide pragmatic information to the compiler). For the case of enumerated types, we need to use Boolean variables to encode the value. This can be done in a fairly simple way by a recursive function that divides the possible values into two subsets of roughly equal size, creates a variable to distinguish between the two, and then recurses on the two subsets. For example, if n is a name, i a natural number, v a symbolic constant, and L

and R are two lists of symbolic constants such that $0 \leq |L| - |R| \leq 1$ then let:

$$\begin{aligned} encode(n.i, v) &= v \\ encode(n.i, (L, R)) &= n.i \rightarrow (encode(n.(i+1), L), encode(n.(i+1), R)) \end{aligned}$$

For example,

$$encode(\text{a}.0, (\text{foo}, \text{bar}, \text{mux})) = \text{a}.0 \rightarrow (\text{a}.1 \rightarrow (\text{foo}, \text{bar}), \text{mux})$$

Certainly other encoding schemes are possible. This one uses the minimal number of boolean variables, $\lceil \log_2 m \rceil$, where m is the number of values.

Using this scheme, the semantics for an enumerated type declaration is

$$[\![\text{VAR } n : (L); \; l]\!] = (n = encode(n.0, L)) \wedge [\![l]\!] \qquad (4.6)$$

That is, for every occurrence of the variable name n, we may substitute the expression yielding the value of n in terms of the encoding variables $n.0, n.1, \ldots$.

Integer subrange variables are handled in exactly the same way, expanding the specified subrange into a list of values. If n is name, x and y are integers, and l is a list of `decl`, then

$$[\![\text{VAR } n : x..y; \; l]\!] = (n = encode(n.0, (x, x+1, \ldots, y))) \wedge [\![l]\!]; \qquad (4.7)$$

4.3.2 Semantics of constants and expressions

In most languages a constant is a constant, but in SMV, this is not so. A constant is an expression, and in SMV, expressions are non-deterministic, meaning they may choose to return one of several possible values. Semantically, we think of an expression as returning a set of values, from which one may be chosen by an `ASSIGN` statement. Thus a constant c is taken as a syntactic abbreviation for $\{c\}$, the singleton containing c. Since an expression is a formula, the simplest way to handle the semantics of expressions is to let an expression denote itself, *ie.*, if e is an `expr`, then

$$[\![e]\!] = e \qquad (4.8)$$

In this case, we require a set of axioms that we can apply to eliminate the various non-Boolean operators from formulas. There are various ways in which this could be done, but one particularly useful way corresponds exactly to the way in which the Ordered Binary Decision Diagram representation is built.

The SMV system

Naturally, this is the way we choose. If o is an operator among the set $+$, $-$, $*$, $/$, mod, $>$, $>=$, $<$, $<=$, $=$, $\&$, $|$, $->$, $<->$, then

$$n = n \to (1,0) \quad \text{where } n \text{ is a variable} \quad (4.9)$$
$$n \to (x,x) = x \quad (4.10)$$
$$(x \to (y,z)) \; o \; a = x \to (\lambda x.(y \; o \; a)1, \lambda x.(z \; o \; a)0) \quad (4.11)$$
$$a \; o \; (x \to (y,z)) = x \to (\lambda x.(a \; o \; y)1, \lambda x.(a \; o \; z)0) \quad (4.12)$$
$$(x \to (y,z)) \; o \; (x \to (a,b)) = x \to (\lambda x.(y \; o \; a)1, \lambda x.(z \; o \; b)0) \quad (4.13)$$

Placing a total order on the variables, and choosing which of the last three axioms to apply based on the order of the leading variables occurring on the left and right of o results in exactly the OBDD form (of course, we only get a savings in time and space if we identify isomorphic terms in our representation and use a cache of previously reduced operators).

The axioms given above can be used to reduce an expression to a form in which the SMV operators occur only in constant subexpressions, which can be reduced to constants by evaluating the base functions. This gives us a computationally effective way to define the semantics of expressions in terms of Boolean formulas. There are certainly other sets of axioms that could be used, however, which might be more suitable to other purposes, such as compiling SMV programs into hardware.

4.3.3 The base functions

The base functions denoted by $+$, $-$, $*$, $/$ are the usual functions of arithmetic modulo 2^{32}. The base function denoted by mod is the positive remainder of division mod 2^{32}. The base functions denoted by the relational operators $>$, $>=$, $<$ and $<=$ are the normal arithmetic comparison operators, and return 0 when the relation is false and 1 when the relation is true. All of these functions are defined for numeric arguments only. For non-numeric values, they return $\{0,1\}$. The equality operator $=$ is defined for all arguments, and returns 0 when they are unequal, and 1 when they are equal. The functions denoted by the Boolean operators are $\&$ (for and), $|$ (for or), $!$ (for not), $->$ (for implies) and $<->$ (for logical equivalence), are defined only for arguments 0 and 1, have their usual truth tables, and return $\{0,1\}$ for other arguments.

When the above base functions are applied to sets, they yield the image of the Cartesian product via the function. That is, if S_1 and S_2 are sets of values,

then
$$f(S_1, S_2) = \{f(x,y) \mid (x,y) \in S_1 \times S_2\}$$

The **union** function takes any two sets and returns their union. The **in** function takes any two sets and returns 1 when the left argument is contained in the right, and 0 otherwise. Within **case** ... **esac** brackets, the syntax $x : y; z$ is an abbreviation for $x \to (y,z)$.

4.3.4 Semantics of ASSIGN

Assigning the value of an expression to a variable in SMV is a little different than in other languages because expresions return sets of possible values. Hence, when we write $v:=e$ in SMV, we really mean $v \in e$, that is, the value assigned to v is chosen among the set returned by e, but is not otherwise determined. There are three cases of ASSIGN, depending on whether the initial, current, or next value of a variable is being assigned. In the first case, if n is a name, e is an expression, and l is a list of decl, then

$$[\![\text{ASSIGN init}(n)\texttt{:=}e;\ l]\!]_I = n \in [\![e]\!] \wedge [\![l]\!] \quad (4.14)$$

Notice here that we are implicitly defining the denotation of a list of declarations to be the conjunction of the denotations of the elements of the list. It may or may not need to be stated explicitly that

$$[\![\text{ASSIGN init}(n)\texttt{:=}e;\ l]\!]_Z = [\![l]\!] \quad (4.15)$$

where Z is one of R, F or S. That is, assignments to the initial value are neutral as regards the transition relation, the fairness constraints or the specification.

For the case of assignment to the current value of a variable, we have:

$$[\![\text{ASSIGN } n\texttt{:=}e;\ l]\!]_R = n \in [\![e]\!] \wedge [\![l]\!] \quad (4.16)$$

The case of assignment to the next value of a variable is complicated by the fact that the assignment only occurs when the process containing it is running. Thus, we have

$$[\![\text{ASSIGN next}(n)\texttt{:=}e;\ l]\!]_R = (n' \in (\text{running} \to ([\![e]\!], n))) \wedge [\![l]\!] \quad (4.17)$$

A conditional asssignment is handled as follows:

$$[\![\text{ASSIGN if } c \text{ then next}(n)\texttt{:=}e;\ l]\!]_R$$
$$=$$
$$(c \Rightarrow n' \in (\text{running} \to ([\![e]\!], n))) \wedge [\![l]\!] \quad (4.18)$$

The SMV system 83

In effect, n' remains undefined when process p is running, allowing p to modify the value of n. We also may need to state that

$$[\text{ASSIGN } n\text{:=}e; \ l]_Z \ = \ [l] \quad (4.19)$$
$$[\text{ASSIGN next}(n)\text{:=}e; \ l]_Z \ = \ [l] \quad (4.20)$$

where Z is one of I, F or S, meaning that these assignments are neutral with regard to the initial condition, the fairness constraints and the specification.

4.3.5 Semantics of DEFINE

The DEFINE construct is similar to ASSIGN except that it assigns a value deterministically. When we write $v\text{:=}e$, we really mean $v = e$. Thus,

$$[\text{DEFINE } n\text{:=}e; \ l]_R = ((n = [e]) \wedge [l]) \quad (4.21)$$

This allows the compiler to use the substitution axiom:

$$(v = e) \wedge f \Rightarrow \lambda v.f(e) \quad (4.22)$$

to eliminate the variable v and treat it instead as a syntactic abbreviation for the expression e.

4.3.6 Semantics of MODULE

A MODULE declaration equates a name with a parameterized list of declarations. Thus, if $n, p_1, p_2, \ldots, p_i$ are names, l is a list of decl and m is a list of module, then

$$[\text{MODULE } n(p_1, p_2, \ldots, p_i) \ l \ m] = (n = \lambda p_1.\lambda p_2 \ldots \lambda p_i.l) \wedge [m] \quad (4.23)$$

To deal with instantiation of modules, it will be useful to introduce more notation in order to construct names in the style of SMV using the "dot" notation. This is most easily accomplished by introducing axioms that allows us to drive the dots inward to the variable names:

$$a.(\neg b) \ = \ \neg(a.b) \quad (4.24)$$
$$a.(b \vee c) \ = \ (a.b) \vee (a.c) \quad (4.25)$$
$$a.c \ = \ c \ \text{ if } c \text{ is a constant} \quad (4.26)$$

We will also need some special brackets to be able to protect certain expressions from the dot operator. For this purpose, we introduce the axiom

$$a.\langle b \rangle = b \qquad (4.27)$$

A module is instantiated by a special form of the VAR declaration. To wit, if k and n are names, $e_1, e_2, \ldots, e_i$ is a list of expr, and l is a list of decl, then

$$[\![\texttt{VAR } k : n(e_1, e_2, \ldots, e_i);\ l]\!] = k.\lambda \texttt{running}.n\langle e_1 \rangle \langle e_2 \rangle \ldots \langle e_i \rangle \langle \texttt{running} \rangle \wedge [\![l]\!] \qquad (4.28)$$

In other words, we obtain the instantiation by substituting the actual parameters $e_1, e_2, \ldots, e_i$ for the formal parameters $p_1, p_2, \ldots, p_i$, and then prefixing the result by the instance name k. Notice that we use angle brackets to protect the arguments from being prefixed by k. In terms of the semantics, this means that the argument expressions are evaluated in the context of the caller rather than the module being called, giving the effect of a call-by-reference. Notice that the variable running is implicitly a parameter of the instance, and that the instance inherits its parent's running variable by virtue of the protecting angle brackets. Hence, the instance runs if and only if its parent runs.

4.3.7 Semantics of process

A *process*, on the other hand, is a module instance which does not inherit its parent's running variable, thought it does inherit certain restrictions on this variable. A process *may* run only if its parent is running, and no more than one process of a given parent may run at any given time. To get this effect in the semantics, we can introduce another variable called blocked. A process, when running, asserts blocked to prevent other processes with the same parent from running. Processes are instantiated thus:

$$\begin{aligned}[\![\texttt{VAR process } k : n(e_1, \ldots, e_i);\ l]\!] &= k.n\langle e_1 \rangle \ldots \langle e_i \rangle \\ &\wedge\ (k.\texttt{running} \Rightarrow \neg \texttt{blocked} \wedge \texttt{running}) \\ &\wedge\ \lambda \texttt{blocked}.[\![l]\!](\texttt{blocked} \vee k.\texttt{running})\end{aligned} \qquad (4.29)$$

By this definition, a process yields to processes defined above it, and takes right of way over processes defined below it. Since the choice of running or not running is non-deterministic, however, all processes of a given parent have the *possibility* of running at any given time. It is also possible that at any given time no processes may be running.

4.3.8 TRANS, INIT, FAIR and SPEC

The SMV system

These four declarations correspond to the four denotations of an SMV program: transition relation, initial conditions, fairness constraints and specification. Their semantics are defined as follows, where e is an expression and l is a list of decl:

$$[\![\texttt{TRANS}\ e;\ l]\!]_T = [\![e]\!] \wedge [\![l]\!] \tag{4.30}$$

$$[\![\texttt{INIT}\ e;\ l]\!]_I = [\![e]\!] \wedge [\![l]\!] \tag{4.31}$$

$$[\![\texttt{FAIR}\ e;\ l]\!]_F = (GF[\![e]\!]) \wedge [\![l]\!] \tag{4.32}$$

$$[\![\texttt{SPEC}\ e;\ l]\!]_S = [\![e]\!] \wedge [\![l]\!] \tag{4.33}$$

Note that the case of the FAIR declaration is slightly different, since the fairness constraint is a conjunction of formulas of the form GFf, meaning that f is true infinitely often, but in the SMV syntax the GF is implicit. As usual, for those denotational categories not defined above, the declarations are semantically neutral.

4.3.9 Semantics of programs

A program is simply a list of module followed by an end-of-file (EOF). Its denotation is that of a parameterless module called main. Thus, if m is a list of module, then

$$[\![m\ \texttt{EOF}]\!] = [\![m]\!] \wedge \texttt{main} \wedge \texttt{running} \tag{4.34}$$

The last conjunct in the denotation gives the effect of a closed universe, by implying that the program is always running (and is never interrupted by an outside process).

The correctness condition for an SMV program p can be stated succinctly as follows:

$$[\![p]\!]_R \models [\![p]\!]_I \wedge [\![p]\!]_F \Rightarrow [\![p]\!]_S \tag{4.35}$$

by which we mean that $[\![p]\!]_I \wedge [\![p]\!]_F \Rightarrow [\![p]\!]_S = $ true in the symbolic model whose transition relation is represented by $[\![p]\!]_R$. This condition is exactly what the SMV model checker computes.

5

A DISTRIBUTED CACHE PROTOCOL

In this chapter, we look at an application of the SMV symbolic model checker to a cache consistency protocol developed at Encore Computer Corporation for their Gigamax distributed multiprocessor [MS91]. This protocol is of interest as a test case for automatic verification for two reasons. First, it is not a theoretical exercise, but a real design, which is driven by considerations of performance and economics, as well as the usual constraints of industrial design, such as compatibility with existing hardware and software. Second, this protocol is a good example of a system where random simulation methods are ineffective in finding all the design errors.

The Gigamax is a distributed, shared memory multiprocessor, in which the processors are grouped into clusters. Each cluster has a local bus, and uses bus snooping [AB86] to maintain cache consistency within the cluster. In addition, each cluster has an interface called a UIC, which links the cluster into a network. The UIC keeps the caches in the cluster consistent with the rest of the network by acting as both a bus snooper and a bus master on behalf of the remote clusters, using a table which keeps track of the remote status of all cache blocks from the local main memory. This allows it to intervene in bus transactions which affect remotely owned blocks, and to send appropriate invalidation or call back requests to the network. The network is organized into a hierarchy, as depicted in figure 5.1. The global bus, at the top of the hierarchy, has one UIC connected to each cluster. These UICs record the status of all cache blocks which are present in the corresponding cluster. This eliminates the need for directory pointers in main memory, at the possible expense of a bottleneck in the global bus.

Protocols such as this are difficult to debug using simulation, in part because

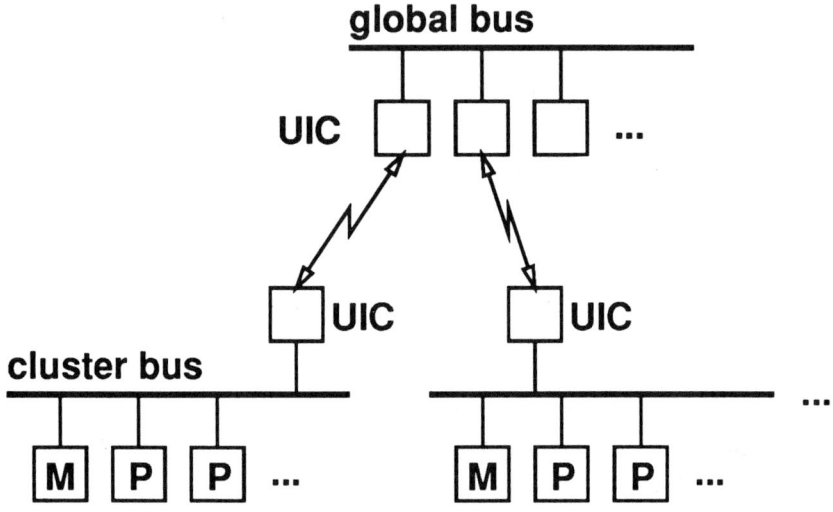

Figure 5.1 Gigamax memory architecture

the order of events such as cache misses and message arrivals in various parts of the system is unpredictable. Subtle errors sometimes require a long sequence of such events to manifest themselves. Since the number of such sequences is combinatoric, the probability of such a sequence occurring in a random simulation rapidly vanishes as the sequence length increases. Nevertheless, for the design process to stabilize, it is necessary to provide timely information about errors to the design team, since the greater the delay in discovering an error, the greater is the disruption required to fix it. Ideally, a protocol should be error free before a hardware (or software) implementation is considered. Otherwise, the options for fixing the errors will be greatly limited by cost considerations, and the likelihood of the design change introducing other errors will be high.

For this reason, we will consider the verification of the Gigamax protocol at a high level of abstraction, neglecting many admittedly important details of the implementation, such as the widespread use of pipelining, or the link level protocol that communicates messages between clusters. The basic method for building an abstract model of a protocol is to introduce *non-determinism* in those cases where the level of detail of the model is insufficient to uniquely determine the outcome of an event, or where design decisions have been left open. We will make a note of places in the model where non-determinism has been used in this way, and in what way the state of an implementation might

A Distributed Cache Protocol   89

correspond to the state of the abstract model.

5.1 THE PROTOCOL

The purpose of a cache consistency system is to provide the illusion to the programmer of a distributed computer that all processors in the system have access to a shared global store. This illusion must be provided despite the fact that the physical storage is distributed. To reduce the latency of access to the distributed main storage, each processor is provided with a local cache – a semi-associative store, which holds a collection of memory blocks recently used by the processor. The time required to access to this store is less than to access main storage. An access to a memory block stored in the cache is called a *hit*, while an access to a memory block *not* stored in the cache is called a *miss*. A miss requires an access to main storage (which may be remote), to retrieve the required memory block and enter it in the cache. This may result in the *replacement* of another block in the cache, to make room for the block being entered in the cache. If the replaced block has been modified while in the cache, it must be returned to main storage. This is called a *copy back* operation.

The first cache consistency protocols for multiprocessors were called *bus snooping* protocols [AB86]. They required that the processors in the system be connected by a bus, or other broadcast medium. In a bus snooping system, each time a memory access occurs over the bus, all of the caches are checked to determine whether they contain the addressed block. If the block is present in a cache, a change in status may be required. For example, if the block is present and modified, the access must be stalled while the modified data are copied back to main storage. In a more sophisticated protocol, the cache with the modified data may supply the data directly to the requesting cache, without the intermediary of main storage. In case of a memory access caused by an attempt to *modify* the data, all caches in which the block is present must *invalidate*, that is, remove the block from cache storage. This insures that all cached copies of the block remain consistent.

The Gigamax protocol uses bus snooping techniques to maintain consistency of the caches within a single cluster. The main difference between the Gigamax snooping protocol and those described in [AB86] is that the Gigamax uses a *split transaction bus*. This means that a processor accessing memory over the bus first places a request on the bus, and then frees the bus for other transactions while awaiting a response. The bus snooping technique is not practical for large scale multiprocessors, because the broadcast medium quickly becomes saturated. For this reason, the Gigamax uses a message passing protocol to maintain consistency between clusters. The split transaction bus protocol allows traffic to continue on the bus while messages are in transit in the network.

The terminology used in the sequel is changed somewhat from the Encore terminology, and the protocol is somewhat simplified to make the presentation clearer. The basic protocol is preserved, however, including a subtle error which was discovered by the SMV system. The following is a description of the protocol, first in English, then in the SMV input language. In the model, we consider only the status of a single memory block. This is our first use of abstraction, and results in nondeterminism in several places in the model.

5.1.1 Processors

Each memory block stored in each cache has an associated *state*, which can be either *invalid*, *shared*, or *owned*. Alternative names for these states would be *absent*, *present*, and *modified*, respectively. The shared state indicates that there may be other processors which have this block stored in their cache. Therefore, a block in the shared state can be read by the processor, but not written, since writing might result in an inconsistency between two caches. The owned state indicates that no other processors have this block in their cache, and that the data in the cache have been modified. Therefore, a block in the owned state can be both read and written by the processor. The invalid state indicates that the block is not present in the cache. Therefore, the block cannot be read or written by the processor.

```
MODULE cache-device

VAR
  state : {invalid,shared,owned};


DEFINE
  readable := ((state = shared) | (state = owned)) & !waiting;
  writable := (state = owned) & !waiting;
```

The split transaction bus snooping protocol works in the following way. At each bus cycle, the bus arbiter chooses a processor among the requesting processors to be the bus *master*. The remaining processors are referred to as *slaves*. The

A Distributed Cache Protocol

master issues a *command* on the bus, of which there are three basic types. A *read* command is a request for a given memory block, and is answered by a *response* command. A *write* command stores data in main memory. The write and response commands can be combined into a single command called a *write-response*, which has the simultaneous effect of supplying data to a requester and storing it in main memory. Each command also signals the next state that the bus master will enter. Thus, a *read-owned* command indicates that the bus master intends to modify the data, and a *read-shared* indicates that it does not. A *write-shared* indicates that the bus master is writing data, but maintaining a shared copy, while a *write-invalid* indicates that it is not keeping the block (*eg.*, it is replacing it with another block). The basic commands, and their uses are summarized in table 5.1. We note that no external command is required to go from the shared state to the invalid state. This occurs when a shared block is removed to make room for another block in the cache. Since our model does not contain the states of any other blocks, we allow this replacement to occur non-deterministically, at any time. Thus we model any possible cache replacement policy.

A slave, observing a command on the bus, may decide to modify its state. For example, a slave observing a read-owned command changes its state to invalid, since the bus master, entering the owned state, will assume it has the only cached copy of the block. Correspondingly, a slave in the owned state observing a read-shared command will change to the shared state. A special command called *invalidate* is used to invalidate all caches in the system. A slave observing this command changes to the invalid state.

```
ASSIGN
  init(state) := invalid;
  next(state) :=
    case
      abort : state;
      master :
        case
          CMD = read-shared          : shared;
          CMD = read-owned           : owned;
          CMD = write-invalid        : invalid;
          CMD = write-shared         : shared;
          1 : state;
        esac;
      !master :
        case
          CMD = read-owned            : invalid;
          CMD = invalidate & !waiting : invalid;
```

from state	command	to state	cause
invalid	read-shared	shared	read miss
invalid or shared	read-owned	owned	write miss
owned	write-invalid	invalid	copy-back
owned	write-resp-invalid	invalid	snoop read-owned
owned	write-shared	shared	write-through
owned	write-resp-shared	invalid	snoop read-shared

Table 5.1 Summary of commands

```
            CMD = read-shared & state = owned : shared;
            state = shared & !waiting : {shared,invalid};
            1 : state;
         esac;
      esac;
```

On receiving the command, each slave checks its own cache and indicates the state of the block in its own cache by asserting the signals *reply-owned*, and *reply-waiting* on the bus. These are *wired or* signals, meaning that the signal is observed to be asserted on the bus if one or more caches assert the signal. The reply-owned signal is asserted by a slave when the block is in the owned state in the slave's cache. Reply-waiting is asserted when the slave has previously requested the block, and is waiting for a response. This signal will be discussed in more detail shortly. The process of looking up the slave's state and signaling on the bus is known as bus snooping. On observing a read command, a slave in the owned state sets a flag called *snoop*. This causes the cache to issue a write-response at a later bus cycle, supplying the data to the requester, and simultaneously storing it in main memory. When this happens, the snoop flag is reset.

An additional reply signal called *reply-stall* may be asserted by any slave, including main storage, if the slave if not ready to respond to the command because some resource is busy. If reply-stall is asserted, the command is nullified.

```
DEFINE
   reply-owned := state = owned;
```

A Distributed Cache Protocol

```
VAR
  snoop : boolean;

ASSIGN
  init(snoop) := 0;
  next(snoop) :=
    case
      abort : snoop;
      state = owned & CMD = read-shared : 1;
      state = owned & CMD = read-owned  : 1;
      CMD = response             : 0;
      CMD = write-resp-invalid   : 0;
      CMD = write-resp-shared    : 0;
      1 : snoop;
    esac;
```

After issuing a read command, the master releases the bus and waits for a *response*. During this time, a flag called *waiting* is set. Normally, if no slave asserts reply-owned, the response comes from main memory. If any slave asserts reply-owned, however, main memory is inhibited, allowing the slave to supply the data at a future cycle with a write-response command.

```
MODULE bus-device

VAR
  master : boolean;
  cmd : {idle,read-shared,read-owned,cty-read,write-invalid,
         write-shared,write-resp-invalid,write-resp-shared,
         invalidate,response};
  waiting : boolean;
  reply-stall : boolean;

ASSIGN
  init(waiting) := 0;
  next(waiting) :=
    case
      abort : waiting;
      master & CMD = read-shared      : 1;
      master & CMD = read-owned       : 1;
      CMD = response                  : 0;
      CMD = write-resp-invalid        : 0;
      CMD = write-resp-shared         : 0;
      1 : waiting;
    esac;
```

A slave which is waiting for a given cache block responds to any read command for that block by asserting reply-waiting. This nullifies the read command and forces the master to retry at a later cycle.

```
DEFINE
  reply-waiting := waiting;
  abort := REPLY-STALL
           | ((CMD = read-shared | CMD = read-owned)
              & REPLY-WAITING);
```

The commands which may be issued by a processor when it is bus master are a function of the state. For example, if the snoop flag is set, the processor may issue a write-response on the bus. From the owned state, a processor may issue a write-invalid command in order to replace the cache block with another. A processor in the shared state may issue a read-owned in case of a write miss, and a processor in the invalid state may issue either a read-shared or a read-owned command, in case of a read miss and write miss respectively.

```
MODULE processor(CMD,REPLY-OWNED,REPLY-WAITING,REPLY-STALL,DATA)
ISA bus-device
ISA cache-device

ASSIGN
  cmd :=
    case
      master & snoop & state = invalid : write-resp-invalid;
      master & snoop & state = shared  : write-resp-shared;
      master & state = owned & !waiting : write-invalid;
      master & state = shared & !waiting : read-owned;
      master & state = invalid : {read-shared,read-owned};
      1 : idle;
    esac;
```

5.1.2 The local UIC interface

The UIC is the interface from one cluster to another. UICs come in pairs, connected by a communication link. A UIC is said to be *local* for a given memory block if that block is found in main storage on the *same side* of the link as the UIC. It is said to be *remote* if the memory block is found in a main

A Distributed Cache Protocol

memory on the *other side* of the link. Thus, for any memory block, one of the UICs in the pair is local, and the other remote. The UIC determines whether it is local or remote by address decoding. We consider the local case first. In this discussion, *local* refers to any part of the system on the bus side of the UIC, and *remote* refers to any part of the system on the link side of the UIC.

Viewed from the bus, the UIC behaves like a processor, with the capability to issue and respond to commands. The UIC's cache records the state (but not the data) of all blocks of the local main storage that are present in remote caches. This allows the UIC to snoop the bus on behalf of remote caches. The UIC performs this function in exactly the same manner as the processors. The state of a block in the UIC changes with commands issued in the same manner as the state of cache blocks in processor caches.

The UIC receives command messages from from the link, and stores them in one of two queues. The low priority queue is for read commands, and the high priority queue is for all other commands. The depth of the queues is arbitrary, but for now, we consider queues of only one entry. A command in one of the queues is issued on the bus when the UIC becomes master. If both queues are non-empty, the command in the high priority queue is issued first. Provided the command is not aborted, the queue issuing the command is emptied. Since the UIC becomes bus master at nondeterministic intervals, the delay between the time a message arrives in the queue and is issued on the bus is arbitrary. This nondeterminism covers two abstractions made in the model. First, it allows for any amount of latency in the link level protocol, which is not modeled. Second, it allows the time to issue an arbitrary number of messages relating to other memory blocks that may be queued ahead of the one message that is modeled.

```
MODULE receiver
VAR
  hiq : {none,response,write-shared,write-resp-shared,
         write-invalid,write-resp-invalid,invalidate};
  loq : {none,read-owned,read-shared,cty-read};

ASSIGN
  cmd :=
    case
      master & !(hiq = none) : hiq;
      master & !(loq = none) : loq;
      1 : idle;
    esac;
  init(hiq) := none;
```

```
  if running then next(hiq) :=
    case
      !master | abort : hiq;
      1 : none;
    esac;
  init(loq) := none;
  if running then next(loq) :=
    case
      !master | abort | !(hiq = none) : loq;
      1 : none;
    esac;
```

The local UIC can send commands to the link in response to commands observed on the local bus. Whenever a read command is sent to the link, it is entered in the remote UIC's low priority queue. If any other command is sent to the link, it is entered in the remote UIC's high priority queue. If the remote queue is full, the local bus cycle is stalled.

```
MODULE sender
DEFINE
  lopri := sending in {read-shared,read-owned,cty-read};
  hipri := sending in {invalidate,response,write-shared,
          write-invalid,write-resp-shared,write-resp-invalid};
ASSIGN
  if !remote.running then next(remote.hiq) :=
    case
      !abort & remote.hiq = none & hipri : sending;
      1 : remote.hiq;
    esac;
  if !remote.running then next(remote.loq) :=
    case
      !abort & remote.loq = none & lopri : sending;
      1 : remote.loq;
    esac;
  reply-stall :=
    (hipri & !(remote.hiq = none) |
     lopri & !(remote.loq = none)) union 1;
```

The local UIC sends command messages to the link in two cases. The first is to invalidate or call back cache blocks in remote caches. This occurs when the UIC is a slave and a read-owned or read-shared is received on the bus. If the UIC is in the owned state, the read-owned or read-shared is forwarded to the link. This causes the remote cache in the owned state to issue a write-resp-invalid or write-resp-shared, returning the cache block to the local bus. If the UIC is in the shared state, and a read-owned is received on the bus, an invalidate command is forwarded to the link. This causes all remote caches to go to the invalid state. Note that this may allow a processor on the local bus to write before the invalidate command has reached all remote caches. This is a possible violation of strict consistency, which is tolerated for performance reasons. Hence, the protocol does not implement a strongly consistent memory model. The memory model which the protocol does support will be discussed in more detail in the next section.

The second case in which the local UIC sends a command to the link is when the UIC has issued a read-shared or read-owned and is waiting for a response. In this case, if the UIC is a slave and a response, write-resp-shared, or write-resp-invalidate is asserted on the bus, a response is sent to the link.

```
MODULE local-UIC(remote,CMD,REPLY-OWNED,REPLY-WAITING,
                REPLY-STALL,DATA)
ISA bus-device
ISA cache-device
ISA receiver
ISA sender

DEFINE
   sending :=
     case
       master : none;
       CMD = read-shared & state = owned  : read-shared;
       CMD = read-owned  & state = owned  : read-owned;
       CMD = read-owned  & state = shared : invalidate;
       CMD = write-resp-invalid & waiting : write-resp-invalid;
       CMD = write-resp-shared  & waiting : write-resp-shared;
       CMD = response           & waiting : response;
       1 : none;
     esac;
```

5.1.3 The Remote UIC interface

When the UIC is remote, it behaves as if it were a main storage device. It accepts read-shared, read-owned, write-shared, and write-invalid commands from the bus, and forwards them to the local UIC via the link. When the response arrives in the high priority queue, it issues the response on the bus. In addition, it can provide a special service to caches on the local side. If the remote UIC issues a read-shared or read-owned command, and there is no reply on the remote bus (*ie.*, no slave asserts reply-owned), it is assumed that the block was copied back to main storage while the read command was in transit. The remote UIC therefore sends the read command back to the local side. This operation is called a *courtesy read*. The courtesy read will cause the main store on the local bus to respond to the original requester.

```
MODULE remote-UIC(remote,CMD,REPLY-OWNED,REPLY-WAITING,
                  REPLY-STALL,DATA)
ISA bus-device
ISA receiver
ISA sender

DEFINE
  sending :=
  case
  master :
    case
    CMD = read-shared & !REPLY-OWNED : cty-read;
    CMD = read-owned  & !REPLY-OWNED : cty-read;
    1 : none;
    esac;
  !master :
    case
    CMD = read-shared & !REPLY-OWNED : read-shared;
    CMD = read-owned  & !REPLY-OWNED : read-owned;
    CMD = write-resp-invalid & waiting : write-resp-invalid;
    CMD = write-resp-shared  & waiting : write-resp-shared;
    CMD = write-resp-shared  & !waiting : write-shared;
    CMD = write-shared   : write-shared;
    CMD = write-invalid  : write-invalid;
    1 : none;
    esac;
  esac;
  reply-owned := 0;
```

The text for the complete model in the SMV language includes such details as an abstracted model of main storage and the cluster bus, which ties the above modules together. These are omitted here. Each cluster is modeled as an asynchronous process. Hence, the early quantification method for disjunctive relations can be used to avoid constructing the global transition relation (cf.

section 3.4.2).

5.1.4 Protocol example

As an example of the protocol in operation, consider the sequence of events depicted in figures 5.2 and 5.3. In the figures, clusters 1 and 2 are both remote (*ie.*, the memory block in question resides in some other cluster). The sequence begins when a read miss occurs in a processor in cluster 2, while a processor in cluster 1 is in the owned state. At this point, the following sequence of events might occur:

1. The processor in cluster 2 issues a read-shared command on the bus, and sets its waiting flag.

2. The UIC in cluster 2 sends the read-shared command up the link, storing it in the low priority queue of the global bus UIC for cluster 2.

3. The global bus UIC for cluster 2 issues the read-shared command on the global bus, entering the shared state, and setting its waiting flag.

4. Since the global bus UIC for cluster 1 is in the owned state, it asserts reply-owned, sends the read-shared command down the link to cluster 1, enters the shared state, and sets its snoop flag.

5. The UIC in cluster 1 issues this read-shared command, entering the shared state and setting its waiting flag.

6. The processor in cluster 1 in the owned state asserts reply-owned, enters the shared state, and sets its snoop flag.

7. The processor in cluster 1 issues a write-resp-shared command, containing the block data, and clears its snoop flag.

8. The UIC in cluster 1 sends the write-resp-shared command up the link, storing it in the high priority queue of of the global bus UIC for cluster 1, and clears its waiting flag.

9. The global bus UIC for cluster 1 issues the write-response-shared command on the global bus, and clears its waiting flag.

10. (a) The global bus UIC connected to main memory sends a write-shared command containing the block data and (b) The global bus UIC for cluster 2 sends a response command, clearing its waiting flag.

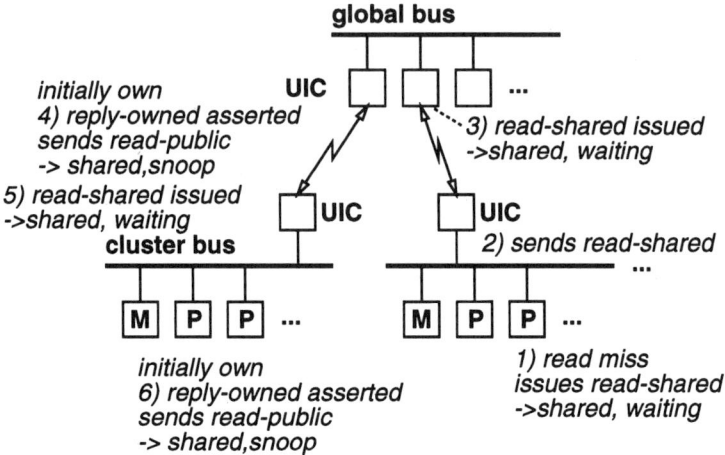

Figure 5.2 Protocol example

11. The UIC in cluster 2 issues the response command.

12. The requesting processor in cluster 2 stores the data in its cache, and clears its waiting flag.

5.2 VERIFYING THE PROTOCOL

We now consider the problem of formal specification and verification of the protocol. The properties we will be concerned with are:

1. freedom from deadlock,

2. sequential consistency, and

3. local safety conditions, related to diagnostics.

Using the symbolic model checking technique, we can verify these properties automatically, despite the very large state space of the model. In fact, the model checker discovered a fairly subtle bug in the protocol – an execution sequence leading to a deadlocked state.

A Distributed Cache Protocol

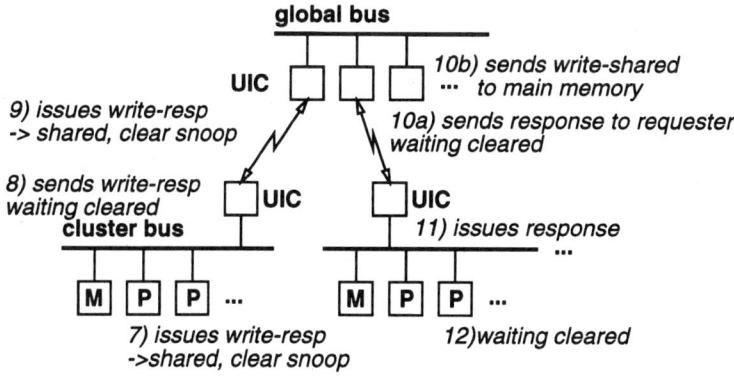

Figure 5.3 Protocol example (cont.)

5.2.1 Freedom from deadlock

We will say that the protocol is deadlocked if it reaches a state in which some processor is permanently blocked from receiving access to the given memory block. Thus, our definition of deadlock takes in situations that might also be called livelock, in which the system continues to loop infinitely, but without the possibility of making progress. We can express this property in CTL with the following formula, which must hold for all processors:

$$AG(EF\, readable \wedge EF\, writable) \tag{5.1}$$

In other words, it it always possible that the memory block will become readable by the given processor, and always possible that it will become writable. We can check this property using SMV by adding the following specification to the processor module:

```
SPEC
   AG(EF readable & EF writable)
```

The specification turns out to be false, and as a counterexample, the model checker produces an execution trace leading to a deadlocked state. This is an actual bug in the original protocol which was found by the model checker, but not in behavioral simulations. The complexity of the counterexample, and the unusual sequence of events that leads to the deadlock should give some indication of why this error would be unlikely to occur in random simulations. The time required to produce the counterexample was slightly under ten minutes running on a Sun 3/60.

The steps of the counterexample are depicted in figures 5.4 to 5.6. Cluster 1 is the local cluster, and clusters 2 and above are remote clusters. We pick up the counterexample at a point where a processor in cluster 2 is in the owned state:

1. A read miss occurs in a processor in cluster 1. This processor issues a read-shared command on the bus. It enters the shared state and sets its waiting flag.

2. Since the UIC in cluster 1 is in the owned state, it asserts reply-owned, enters the shared state, and sends a read-shared command up the link, storing it in the low priority queue of the global bus UIC for cluster 1.

3. A processor in cluster 3 also issues a read-shared command. As a result, the global bus UIC for cluster 3 issues the read-shared command on the global bus, entering the shared state, and setting its waiting flag.

4. Since the global bus UIC for cluster 2 is in the owned state, it asserts reply-owned, sends a read-shared command down the link to cluster 2, enters the shared state, and sets its snoop flag.

5. The UIC in cluster 2 issues this read-shared command, entering the shared state and setting its waiting flag.

6. The processor in cluster 2 in the owned state asserts reply-owned, enters the shared state, and sets its snoop flag.

7. The processor in cluster 2 issues a write-resp-shared command, containing the block data, and clears its snoop flag.

8. The UIC in cluster 2 sends the write-resp-shared command up the link, storing it in the high priority queue of of the global bus UIC for cluster 1, and clears its waiting flag.

9. The global bus UIC for cluster 2 issues the write-response-shared command on the global bus, and clears its waiting flag.

10. (a) The global bus UIC connected to main memory (cluster 1) sends a write-shared command containing the block data and (b) The global bus UIC for cluster 3 sends a response command, clearing its waiting flag.

11. The UIC in cluster 1 issues the write-shared command.

A Distributed Cache Protocol

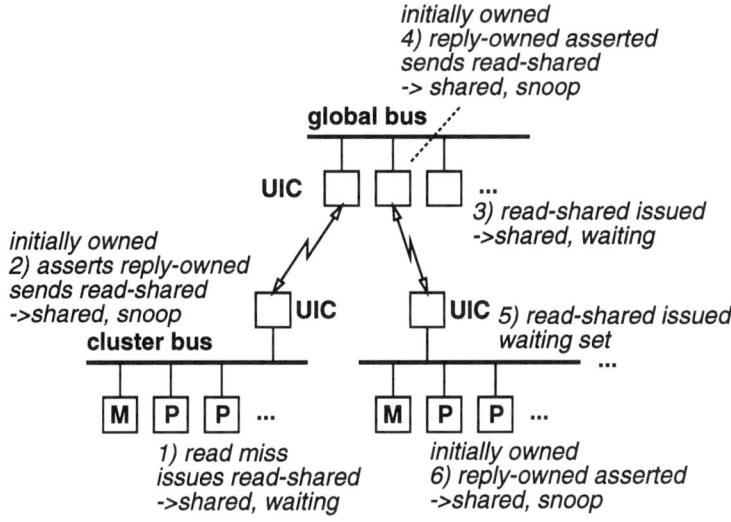

Figure 5.4 Deadlock example

12. The block data are stored in main memory.

13. A processor in cluster 3 again issues a read-shared command. As a result, the global bus UIC for cluster 3 issues the read-shared command on the global bus, entering the shared state, and setting its waiting flag.

14. Since read-owned is not asserted, the UIC for cluster 1 sends the read-shared command down the link towards main memory.

At this point, the system is deadlocked. The original read-shared command sent in step 1 in cluster 1 is still in the low priority queue at th global bus level, but is stalled by the waiting flag set in the global UIC for cluster 3. Similarly, the read-shared command sent by cluster 3 is in the low priority queue in the cluster 1 UIC, but is stalled by the waiting flag of the original requester. This is an example of the classic deadlock situation which occurs when two processes attempt to obtain locks on two resources (in this case two buses) in different orders. Nonetheless, the sequence of events that lead to this situation were sufficiently complex that the designers did not anticipate that the situation could occur, and simulations did not produce it. In fact, the deadlock situation was found at a search depth of thirteen transitions. At each step in this sequence, there were several alternatives that might have averted the

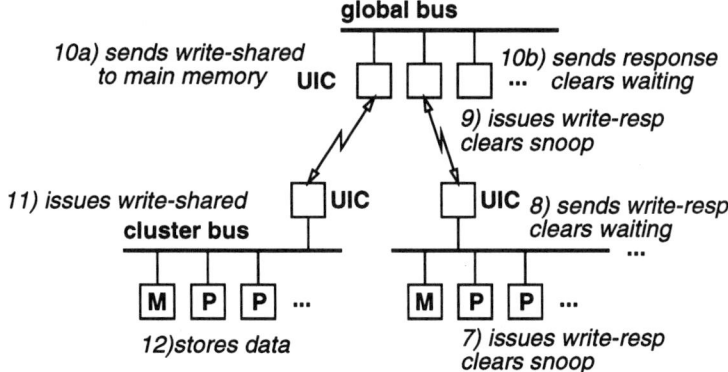

Figure 5.5 Deadlock example (cont.)

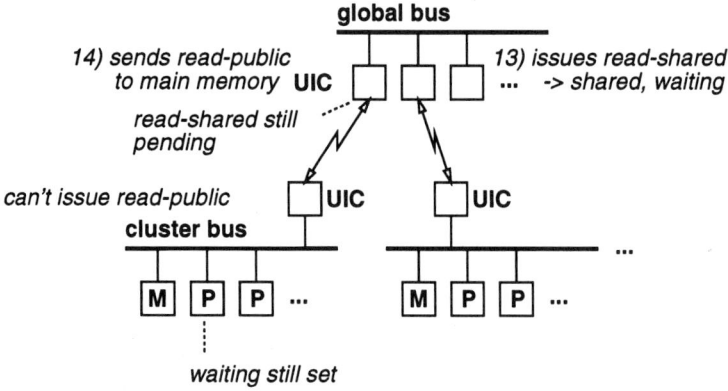

Figure 5.6 Deadlock example (cont.)

A Distributed Cache Protocol   105

deadlock. Thus it is possible, but unlikely that this deadlock would be found by a random simulation run, or a simulation run based on address traces.[1]

The fact the the model checker was able to print out automatically an example of this deadlock highlights an important practical aspect of the technique. Counterexamples are of perhaps even greater value than a proof that the system is correct, since such a proof is based on the assumption that the system is correctly modeled, and the specification is correct and complete. A counterexample, however, provides an important clue as to where a bug in the system lies, and how it might be corrected.

5.2.2 Correcting the deadlock

The problem causing the deadlock is that the remote owner of the memory block can write the data back to main memory while a read command from the local cluster is in transit to the remote cluster. The write command crosses the read command in the mail, so to speak. A remote request for the same block can then lock the global bus, leading to deadlock. The Encore engineers corrected the deadlock problem in the following way. The write command, when it reaches the local bus, is converted by the UIC into a write-response command. This supplies data to the local requester and frees the local bus. Unfortunately, it also leaves an orphan read command in the system. If a read command from the remote side is issued on the local bus, and a remote processor subsequently reaches the owned state, the orphan read command will disrupt th protocol. To prevent this, when the orphan read is issued, it is converted to a special command called *echo-response*, which is sent back to the local cluster. The UIC in the local cluster stalls any commands on the local bus until the echo-response arrives, thus guaranteeing that the orphan read command is destroyed.

The corrected model satisfies the absence of deadlock specification. The performance of the SMV model checker in verifying this is plotted in figure 5.7, for a model with 2 clusters, as the number of caches in each cluster is increased from 2 to 6 (thus, in the largest model, there are 12 caches and 4 UICs). Part (a) shows the run time as a function of the number of caches per cluster. Part (b) shows the number of OBDD nodes used overall, and for representing

---

[1] In fact, the number of possible transitions from a given state ranges from 6 to 12. The probability of a random simulation run executing this trace is therefore in the range $6^{-13} = 7.7 \times 10^{-11}$ to $12^{-13} = 9.3 \times 10^{-15}$. The expected time for a random simulation to exhibit this behavior would be somewhere between 2.4 years and 29 millenia, assuming the simulation could be carried out at 10,000 steps per second.

the transition relation. Part (c) shows the number of reachable states of the model. Although the run time points are well fit by a quadratic curve, the actual asymptotic performance is most likely cubic, as in the case of the synchronous arbiter (cf. section 3.4.1), owing to linear increases in the transition relation size, the number of fixed point iterations and the size of the OBDDs representing fixed point approximations.

Since the number of bus wires running between successive caches is fixed, we can apply theorem 4 to show that the transition relation OBDD size must grow linearly in the number of caches. The fact that the fixed point approximation OBDDs also grow linearly bears further examination, however. This phenomenon can be understood by considering the nature of the protocol. Imagine cutting a cluster bus in half, and consider how much information must be communicated from one half of the bus to determine whether a given state of the system is in the reachable set or not. In fact, this amount is fixed, independent of the number of caches on the bus, since we need only know if there are any caches in the shared state or the owned state on the other side of the cut, and not in particular which caches these are or how many. As a result, the number of OBDD nodes (representing the reached state set) at the level corresponding to our cut is bounded.[2] This is characteristic of bus snooping protocols, and other protocols which are "loosely coupled", in the sense that one half of the system has bounded knowledge of the state of the other half of the system.

As part (c) of the figure shows, the number of states of the system increases exponentially with the number of caches per cluster. Despite this, the performance of the symbolic model checking algorithm is polynomial. Thus, for this particular model and specification, we have solved the state explosion problem.

5.2.3 Sequential consistency

When writing a formal specification for the Gigamax cache consistency protocol, we need to consider the model of a distributed memory which the Gigamax provides to the programmer. As mentioned previously, for performance reasons the protocol does not maintain strict consistency of the caches. A cache block in the shared state may be out of date for a short time while an invalidate message is traversing the network. This is tolerated, since maintaining strict consistency would require an acknowledgment of invalidation to be collected from all caches in the shared state before a cache block could be modified.

---

[2] For other applications of this kind of argument, see [Bry91].

A Distributed Cache Protocol

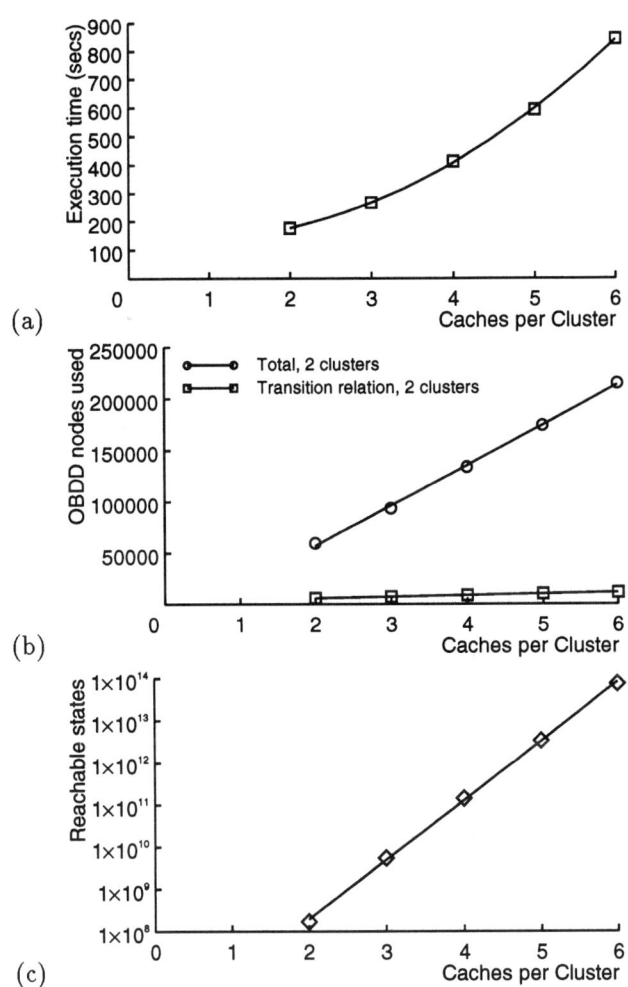

Figure 5.7 Performance for checking deadlock

There are a number of distributed memory models that may be supported by such a system. A *totally ordered model* is one in which all processors observe all values written to the memory in the same order. For example, in a totally ordered model, if the processors write into a location the sequence of values $1, 2, 3, \ldots$, then all processors which read the location will observe any new values to be greater than or equal to all previous values. We will show that the Gigamax protocol has this property, for a one block system. In a partially ordered model, values written may in some cases be observed in a different order by different processors. Some guarantee of ordering is usually made. For example, all writes must be observed in the same order relative to special synchronization operations, or all writes by the same processor must be observed in the same order. The latter model is supported by the Gigamax protocol for writes to different cache blocks. Since our model of the protocol only describes the behavior one cache block, however, the model cannot be used to check this property.

Returning to the problem of total ordering of writes to the same block, it might seem at first that there is no "finite state" description of a protocol that writes an unbounded sequence of values. We can check the property, however, by using an abstraction. We do this by choosing a value n, and storing in the model only one bit of information – whether the data value is less than n or greater than or equal to n. We then assume that the processors never write a value less than n after a value greater than n has been written, and we show that a processor never reads a value less than n after reading a value greater than n. Since the value of n is arbitrary, it follows that all processors read data values in non-decreasing order, satisfying the total ordering requirement. We now consider how to model the system using this abstraction. For each cache, we introduce a variable whose value is 0 when the data value is less than n and 1 when the date value is greater than or equal to n. This variable may change whenever the block is writable, but may only change from 0 to 1, since we assume the processors only increase the data value. The following SMV code models the data held in the processor's cache:

```
MODULE data-device
VAR
   data : boolean;
ASSIGN
   next(data) :=
     case
        !master & waiting & CMD in {response,write-resp-invalid,
                  write-resp-shared} : DATA;
        writable : data union 1;
```

A Distributed Cache Protocol

```
        1 : data;
    esac;
DEFINE
    data-enable := master & CMD in {response,write-resp-invalid,
                    write-resp-shared,write-invalid};
```

Additionally, we introduce variables to represent the values on the buses and the values in the high priority message queues. The low priority queues hold only requests, which have no data value.

We would now like to prove that this abstract model of the data path of the protocol satisfies the following specification in CTL, for all processors:

$$AG[(readable \wedge data \geq n) \Rightarrow AG\neg(readable \wedge data < n)] \quad (5.2)$$

In other words, if ever a value greater or equal to n is observed, a value less than n is never observed in the future. We can check this using SMV by adding the following specification to the processor module:

```
SPEC
    AG(readable & data -> AG (readable -> data))
```

Figure 5.8 shows the performance of the symbolic model checking algorithm in verifying this formula, again for a model with 2 clusters. Part (a) of the figure shows the execution time, while part (b) shows the amount of storage used. Notice that although the execution times are roughly ten times those obtained for the model without data, they are still cubic in the number of processors per cluster.

5.2.4 Correctness of diagnostics

In addition to the above specifications, it was also particularly useful to check that the diagnostics built into the protocol never flagged an error under normal operation of the protocol. Errors are flagged by the diagnostic system in each processor subsystem whenever a command is observed on the bus which is inconsistent with the processor's local state. Determining which command/state combinations are normal, and which are errors is difficult, and a number of errors of this type were found in the protocol using the model checking technique.

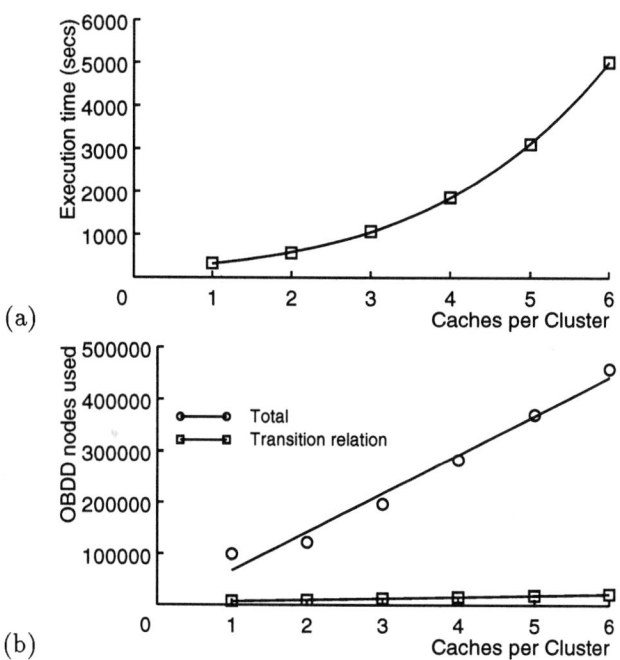

Figure 5.8 Performance for checking sequential consistency

5.3 DISCUSSION

In verifying the Gigamax model with respect to the formal specifications, the symbolic model checker was able to perform an exhaustive search of the model's state space without explicitly constructing the global state graph. As a result, the state explosion problem was avoided. In addition, the model checker exposed a number of subtle errors in the design that were not found in simulation. These errors were usually caused by events (*eg.*, cache misses and message arrivals) occurring out of the normal sequence anticipated by the designers. This type of error is difficult to find in random simulations, since the probability of a given sequence of random events occurring by pure chance is in inverse exponential proportion to the length of the sequence. As we have seen, the sequences necessary to produce protocol errors can be quite long. As the design evolved to correct the errors found by model checking, the model was easily adapted, and quickly provided an analysis of any new errors introduced by design changes. This tends to amortize the initial effort required to produce the protocol model. The ability of the symbolic model checker to find errors quickly makes it easier to experiment with alternative designs, and also helps to build the designer's intuition about the behavior of the system. This is important, because designers tend to concentrate on normal sequences of events, and overlook the unusual sequences. The use of OBDDs in the symbolic model checker made it possible to check a model that would have been very time consuming, or perhaps impossible to check using earlier algorithms.

At this point, the technique has a number of limitations. One limitation is the use of OBDDs. For example, while we find the OBDD sizes growing polynomially in the number of caches in the Gigamax model, if we instead increase the number of cache blocks and leave the number of caches constant, we find the size of the OBDDs increasing exponentially. As a result, it was extremely difficult to check specifications of a system with just two cache blocks (some runs took up to a week, and others never finished). In these cases, the size of the OBDDs representing the fixed point approximations became intractably large. When this happens, techniques such as early quantification that make the representation of the transition relation smaller are little use, since they do not effect the size of the OBDDs representing fixed point approximations.

Another major issue is implementation of the protocol. Clearly, verification of the protocol itself is important, since a correct protocol is a prerequisite for a correct implementation. This is, of course, only half the story. Techniques are also needed to insure that the verified protocol is implemented correctly in hardware. This can in fact be done, using a process of successive refinement of finite state systems that has been studied extensively by Kurshan [Kur87]. The work of Bose and Fisher [BF89a] is also an example of this. Unfortunately, the truth of CTL formulas containing existential quantifiers is not necessarily preserved by this kind of refinement. Thus, for example, though the high level protocol may be deadlock free, a specific implementation of the protocol may not be deadlock free[3]. In order for the implementation to preserve all CTL properties of the protocol, the two would have to be bisimular (cf. section 6.4). Since this is a very strong requirement, it is not clear that the protocol could in fact be implemented with this degree of accuracy in an efficient way. For essentially this reason, Grumberg and Long have studied the use of a subset of CTL using only universal path quantifiers for hierarchical reasoning [GL91]. In any event, though checking the absence of deadlock specification was very useful in finding bugs in the protocol, we must attach a special caveat to this result, since it does not guarantee that all reasonable implementations of protocol will be deadlock free.

Finally, there is the problem of verifying a model with a finite number of processors, when there is no finite limit on the number of processors that could in principle be added to the system. In practice, the intended maximum number of processors is approximately 100. Even using the symbolic model checking technique, however, checking a system of 100 processors seems infeasible at present, and 1000 processors is out of the question. To deal with systems with a very large number of identical components, we can apply methods of induction over processes. As in the case of successive refinement, induction methods are not fully automatic – some human input is required in the form of an inductive hypothesis. In chapter 7, we will deal with the problem of induction over processes.

---

[3] In fact, such a deadlock, involving an interaction between the memory and processor subsystems was known to the Encore engineers. The memory system, when busy, would stall any new requests, but the stalled request would still remain in the memory system's pipeline for four clock cycles. Thus, when the processor retried the request four clock cycles later, it would be stalled again, and the process would repeat indefinitely.

6

MU-CALCULUS MODEL CHECKING

The Mu-Calculus [Par74] is a logic based on extremal fixed points that is strictly more expressive than CTL,[1] and can also express a variety of properties of transition systems, such as reachable state sets, state equivalence relations, and language containment between automata. A symbolic model checking algorithm for this logic allows all of these properties to be computed using OBDDs [BCM+90].

6.1 THE Mu-Calculus

We have already seen a number of Mu-Calculus formulas. For example, in section 3.2, we characterized the CTL formula EXp as

$$\lambda x.\exists y.(R(x,y) \wedge p(y))$$

That is, EXp is the set of states x such that there exists a state y such that (x, y) is in the transition relation R and y is in the set of states satisfying p. As another example, we characterized EFp as

$$\mu y.(p \vee EXy)$$

That is, EFp is the least fixed point of a functional $\tau = \lambda y.(p \vee EXy)$. This functional takes a set of states y and returns a set of states $(p \vee EXy)$. In general, there is a least fixed point (and a greatest fixed point) of such a functional

---

[1] Emerson and Lei [EL86] gave a model checking algorithm for a somewhat different version of the Mu-Calculus, and showed that there are formulas in this logic that cannot be expressed in CTL. Here, we use the relational Mu-Calculus of Park [Par74]

if it is *formally monotonic*, meaning that the parameter y occurs only within an even number of negations. This is a sufficient (but not necessary) condition for the functional to be monotonic.

Since a wide variety of verification conditions (other than satisfaction of a CTL formula by a model) can be expressed using fixed points, it is useful to formalize a model theory for this calculus, and to develop a symbolic model checking technique for it, much in the same way we did for CTL. In this way, we have only to express the desired conditions in the Mu-Calculus, and we have a ready made symbolic algorithm for checking them.

We start with an exact definition of the Mu-Calculus formulas. There are two kinds of formulas, which we will refer to as *relational formulas* and *Boolean formulas*. There are also two kinds of variables, *relational variables* (such as R, the transition relation) and *individual variables* (such as x, a state). A model for the Mu-Calculus is a triple $M = (S, \phi, \psi)$, where S is a set of states, ϕ is the *individual interpretation*, which maps every individual variable onto an element of S, and ψ is the *relational interpretation* which maps every n-ary relational variable onto subset of S^n. The transition relation R, for example, is a binary relational variable.

The syntax of Boolean formulas is defined as follows:

1. true and false are Boolean formulas.
2. If p and q are Boolean formulas, then so are $p \vee q$ and $\neg p$.
3. If p is a Boolean formula and x is an individual variable, then $\exists x.p$ is a Boolean formula.
4. If R is an n-ary relational formula, and $(x_1, \ldots, x_n)$ is a vector of individual variables, then $R(x_1, \ldots, x_n)$ is a Boolean formula.

The formula $\exists x.p$ is true when there exists a state x in S such that p is true. The formula $R(x, y)$ is true if the pair $(\phi(x), \phi(y))$ is in the relation $\psi(R)$.

The relational formulas are defined as follows:

1. Every n-ary relational variable R is an n-ary relational formula.
2. If p is a Boolean formula and $(x_1, \ldots, x_n)$ is an n-tuple of individual variables, then $\lambda(x_1, \ldots, x_n).p$ is an n-ary relational formula.

Mu-Calculus model checking

3. If R is an n-ary relational variable and F is an n-ary relational formula that is formally monotonic in R, then $\mu R.F$ and $\nu R.F$ are relational formulas.

In a given model (S, ϕ, ψ), we identify the relational variable R with the relation $\psi(R)$. The formula $\lambda(x_1, \ldots, x_n).p$ stands for the set of all n-tuples $(x_1, \ldots, x_n)$ such that p is true. The formula $\mu R.F$ stands for the least fixed point of $\tau = \lambda R.F$, while $\nu R.F$ stands for the greatest fixed point.

6.2 SYMBOLIC MODELS

Now, given the above semantics for Mu-Calculus formulas with respect to models, we would like to formulate a symbolic representation for models using Boolean formulas, and to characterize the various operators with repect to this representation. As in the case of Kripke models, we will assume that the set of states S is the set of Boolean vectors $\{\text{true}, \text{false}\}^n$. As in CTL model checking, we can represent a set of states p (a unary relation) with a Boolean **p** formula according to:

$$p = \lambda(v_1, \ldots, v_n).\mathbf{p} \tag{6.1}$$

The choice of variables $v_1, \ldots, v_n$ fixes the representation. Similarly a binary relation R is represnted by a formula **R** according to:

$$R = \lambda(\vec{v}_1, \vec{v}_2).\mathbf{R} \tag{6.2}$$

where $\vec{v}_1$ and $\vec{v}_2$ are n-tuples of Boolean variables, and in general an m-ary relation can be represented thus:

$$R = \lambda(\vec{v}_1, \ldots, \vec{v}_m).\mathbf{R} \tag{6.3}$$

where $\vec{v}_1, \ldots, \vec{v}_m$ are n-tuples of Boolean variables.

Given a relational interpretation ψ (which we can now represent symbolically), let us identify every Boolean formula in the Mu-Calculus with the set of individual interpretations ϕ such that the formula is true. As usual, we identify $p \wedge q$ with the intersection of p and q, $p \vee q$ with the union of p and q, and $\neg p$ with the complement of p. Let the individual variables be $x_1, \ldots, x_k$. An indivudual interpretation ϕ can then be treated as a vector $(\phi(x_1), \ldots, \phi(x_n))$. Now, for each individual variable x_i, let us choose a vector of Boolean variables $\vec{x}_i = (x_{i1}, \ldots, x_{in})$. We can now represent any Boolean formula f in the Mu-Calculus with an ordinary Boolean formula **f** such that

$$f = \lambda(\vec{x}_1, \ldots, \vec{x}_k).\mathbf{f} \tag{6.4}$$

6.3 SYMBOLIC ALGORITHM

To develop a symbolic model checking algorithm for the Mu-Calculus, we need to characterize the operators of the logic with respect to our symbolic representation. To begin with, suppose we have a Boolean formula $R(x,y)$, where R is a relational variable and x and y are individuals. This is the set of all individual interpretations such that (x,y) is in the relation R, or

$$R(x,y) = \lambda(\vec{x}_1,\ldots,\vec{x}_k).\psi(R)(x,y)$$

Now in our symbolic model $\phi(R) = \lambda(\vec{v}_1,\vec{v}_2).\mathbf{R}$, so, using equation 6.4 for the symbolic representation of a Boolean Mu-Calculus formula, we have

$$\mathbf{R}(x,y) = (\lambda(\vec{v}_1,\vec{v}_2).\mathbf{R})(\vec{x},\vec{y})$$

where $\vec{x}$ and $\vec{y}$ are the vectors of Boolean variables thet we chose to represent the individual variables x and y. In other words, we get the representation for $R(x,y)$ by substituting $\vec{x}$ and $\vec{y}$ for $\vec{v}_1$ and $\vec{v}_2$ in the symbolic representation for R. If we are using OBDD's as our representation, this can be done using the *Compose* algorithm (see section 3.3).

The Boolean operators of the Mu-Calculus correspond exactly to their ordinary Boolean counterparts on the symbolic representation. For example, the Boolean Mu-Calculus formula $\exists x.f$ is the set of individual interpretations such that f holds for some value of x. In the symbolic representation,

$$\exists x.\mathbf{f} = \exists \vec{x}.\mathbf{f}$$

where $\vec{x}$ is the vector of Boolean variables representing the individual variable x.

Now suppose we have a relational formula $\lambda(x,y).f$ where f is a Boolean formula. This is the set of state pairs (x,y) in S^2 such that f holds. In fact, we need to be careful here, since $\lambda(x,y).f$ only represents a set of pairs of states when it is closed, *ie.*, when it returns true or false for any pair (x,y). In general, the truth value might depend on the interpretation of the individual variables. To avoid this case, we could require that formulas of this type be syntactically closed, in the sense that only variables x and y may occur free in f. On the other hand, we could say that a relational term applied to the pair (x,y) identifies a set of individual interpretations in which the relation holds. In either event, the symbolic representation is given by

$$\lambda(x,y)\mathbf{f} = (\lambda(\vec{x},\vec{y}).\mathbf{F})(\vec{v}_1,\vec{v}_2)$$

In other words, we obtain the representation for the relation $\lambda(x,y).f$ by substituting the Boolean variable vectors $\vec{v}_1$ and $\vec{v}_2$ for $\vec{x}$ and $\vec{y}$ in the representation for f. Again, this can be accomplished for OBDD's with the *Compose* algorithm. Note that if $\lambda(x,y).f$ is closed, we will obtain a representation of a pure relation, otherwise we will obtain a hybrid representation corresponding to the set of individual interpretations $(\vec{x}_1, \ldots, \vec{x}_k)$ for which $(\vec{v_1}, \vec{v_2})$ is in the relation. In this case, the Boolean variables representing individual variables (the $\vec{x}_i$) and representing relational parameters (the $\vec{v}_i$) must be distinct.

Finally, we come to fixed point formulas, $\mu y.f$ and $\nu y.\mathrm{f}$, where f is a relational formula and y is a relational variable. We have assumed that f is formally monotonic, meaning that y occurs in f only inside an even number of negations. From this, we can show that $\tau = \lambda y.f$ is monotonic, both in our standard semantics and our symbolic representation. The proof of this is straightforward by induction over the structure of formulas, but we omit it here. Since τ is monotonic, and our state space S is finite, the extremal fixed points $\mu y.f$ and $\nu y.f$ are reached in a finite number of iterations of τ on false and true respectively.

This gives us the algorithm for evaluating Mu-Calculus formulas which is described in figure 6.1. Note that the number of iterations of a functional required to reach a fixed point is at most $|S| = 2^n$. However, since fixed point operators may be nested, each iteration may itself involve evaluating a fixed point series. In general, this algorithm evaluates $O(|S|^d)$ formulas, where d is the nesting depth of μ and ν.

6.4 APPLICATIONS OF THE Mu-Calculus

The Mu-Calculus is quite expressive, as can be seen by the following compendium of applications. To begin with, given a binary relation R, the image of a set Q via R is

$$R(Q) = \lambda y.\ \exists x.\ (R(x,y) \wedge Q(x))$$

The set reachable from Q in any number of steps of R (including 0) is

$$R^*(Q) = \mu Y.\ (Q \vee R(Y))$$

The transitive (irreflexive) closure of the relation R is

$$R^+ = \mu Y.\ \lambda(x,z).\ [R(x,z) \vee \exists y.\ (Y(x,y) \wedge Y(y,z))]$$

function eval(f, ψ)
    case
        $f = R(x_1, \ldots, x_n)$: **return**$((\lambda(\vec{v}_1, \ldots, \vec{v}_n).\psi(R))(\vec{x}_1, \ldots, \vec{x}_n))$
        $f = \neg p$: **return** $\neg$eval(p, ψ)
        $f = p \vee q$: **return** eval$(p, \psi) \vee$ eval(q, ψ)
        $f = \exists x.\ p$: **return** $\exists \vec{x}$. eval(p, ψ)
        $f = \lambda(x_1, \ldots, x_n).p$: **return**$((\lambda(\vec{x}_1, \ldots, \vec{x}_n).\text{eval}(p,\psi))(\vec{v}_1, \ldots, \vec{v}_n))$
        $f = \mu Y.\ p$: **return** fixedpoint$(Y, p, \psi\langle Y \leftarrow \text{false}\rangle)$
        $p = \nu Y.\ p$: **return** fixedpoint$(Y, p, \psi\langle Y \leftarrow \text{true}\rangle)$
    end case
end function

function fixedpoint(Y, p, ψ)
    Y' = eval(p, ψ)
    if Y' = $\psi(Y)$ **then return Y'**
    else return fixedpoint$(Y, p, \psi\langle Y \leftarrow \mathbf{Y'}\rangle)$
end function

Figure 6.1 Symbolic Mu-Calculus model checking algorithm

CTL and fairness constraints

The interpretation of the operators of CTL in a Kripke model (S, R, L) can be characterized in the Mu-Calculus as follows:

$$\begin{aligned} EXp &= \lambda x.\ \exists y.\ (R(x,y) \wedge p(y)) \\ EFp &= \mu y.\ (p \vee EXy) \\ EGp &= \nu y.\ (p \wedge EXy) \\ E(q\ U\ p) &= \mu y.\ (p \vee (q \wedge EXy)) \end{aligned}$$

In addition to these standard operators, we can also characterize the CTL operators under *fairness constraints*. A fairness constraint in its simplest form is a condition GFf, meaning f holds infinitely often, which is assumed to be true along all computation paths. Such conditions can be used to enforce fair scheduling of processes and access to resources. They are not directly expressible in CTL, since the tense operators F and G cannot be directly combined. Instead, we restrict the path quantifiers of CTL to apply only to those paths along which each formula in a set C holds infinitely often. To distinguish these constrained path quantifiers from ordinary path quantifiers, we subscript them with C. Thus, $A_C f$, where C is a set of CTL formulas and f is a linear formula, means that for all paths, if each formula of C is true infinitely often, then f is true. Similarly, the formula $E_C f$ means that there exists a path such that each formula of C is true infinitely often and f is true. Here, we consider only the CTL operators with existential path quantifiers, since the operators with universal quantifiers can be derived from these.

The formula $E_C G p$ is true when there is some path in which p is true in every state, and each element of C is true infinitely often. Let

$$\tau = \lambda y. p \wedge EX \bigwedge_{c \in C} E(y \; U \; (y \wedge c)).$$

We argue as follows that $E_C G p$ is the greatest fixed point of τ. First, if y is a fixed point, then every state in y satisfies p, and further, has a nontrivial path remaining in y which leads to a state satisfying each fairness constraint. Hence, a looping path can be constructed satisfying each infinitely often without exiting y. Thus $y \subseteq E_C G p$. On the other hand, suppose $y = E_C G p$. Since every state in y has a path touching each fairness constraint infinitely, as does each state along that path, it follows that every state in y can reach every fairness constraint without exiting y. Thus $y \subseteq \tau(y)$. Therefore, $E_C G p$ is the greatest fixed point of τ. The set of states satisfying $E_C G p$ is expressed in the Mu-Calculus as

$$\nu y. \; (p \wedge EX \bigwedge_{c \in C} E(y \; U \; (y \wedge c)))$$

The remaining operators under fairness constraints can be characterized in terms of $E_C G p$, as follows:

$$\begin{aligned} E_C X p &= EX(p \wedge E_C G \text{ true}) \\ E_C F p &= EF(p \wedge E_C G \text{ true}) \\ E_C(q \; U \; p) &= E(q \; U \; (p \wedge E_C G \text{ true})) \end{aligned}$$

Emerson and Lei [EL86] give a characterization in the Mu-Calculus of CTL under a more general class of fairness constraints. Each constraint in this scheme requires that one condition holds infinitely often or a second condition holds finitely often (for example, either acknowledge holds infinitely often, or request holds finitely often).

Simulation relations

Two states x and y of a Kripke model are said to be *bisimular* if:

1. x and y agree on the atomic propositions,

2. every successor of x is bisimular to a successor of y and

3. every successor of y is bisimular to a successor of x.

Two states are bisimular if and only if they satisfy the same set of CTL formulas [BCG87]. If $(a_1, a_2, \ldots, a_k)$ are the atomic propositions, then the bisimulation relation can be expressed in the Mu-Calculus as follows:

$$\begin{aligned} Bisim \ = \ &\nu Y.\ \lambda x, y.\ \bigwedge_{1 \leq i \leq k} (a_i(x) \iff a_i(y)) \\ &\wedge \forall x'.\ (R(x, x') \Rightarrow \exists y'.(R(y, y') \wedge Y(x', y'))) \\ &\wedge \forall y'.\ (R(y, y') \Rightarrow \exists x'.(R(x, x') \wedge Y(x', y')))) \end{aligned}$$

where we have, as usual, identified each atomic proposition with the set of states in which it is true. There is also an asymmetric notion of simulation – we say that a state x simulates a state y if:

1. x and y agree on the atomic propositions,

2. every successor of y is simulated by a successor of x.

If state x simulates state y, then y satisfies every formula satisfied by x in a dialect of CTL called ∀-CTL, which allows only universal path quantifiers [GL91].[2] Testing bisimulation and simulation relations can be used as a form of verification, or it can be used to test abstractions used in compositional model checking techniques [CLM89a, GL91]. The same idea can easily be extended to models with labeled transitions.

Language containment

The Mu-Calculus can can express the relation of language containment between two deterministic ω-automata. For the sake of simplicity, we consider only deterministic Büchi automata, which are not complete for the class of ω-regular languages, but it is not substantially more difficult to handle more general classes of deterministic automata, such as Street automata.

---

[2] In fact, this is also true for CTL*, an extension of CTL which allows unrestricted linear temporal formulas to be preceded by path quantifiers.

Mu-Calculus model checking

A finite deterministic Büchi automaton consists of a set of states K, an initial state $p_0 \in K$, an alphabet Σ, a set of transitions $\Delta \subseteq K \times \Sigma \times K$, and an acceptance set $B \subseteq K$. The transition relation is such that, for any state p and symbol σ, there is exactly one q for which $\Delta(p, \sigma, q)$. The automaton accepts an infinite sequence $\sigma \in \Sigma^\omega$ iff the sequence of states p, where $\Delta(p_i, \sigma_i, p_{i+1})$ holds for all i, passes through the acceptance set B infinitely often. The set of sequences accepted by an automaton M is called the language of M and denoted $\mathcal{L}(M)$.

To determine whether the language of a Büchi automaton M is contained in the language of a Büchi automaton M' (with the same alphabet), we define a Kripke structure representing the product of M and M', and write a formula in CTL which is true if and only if every sequence accepted by M is also accepted by M' [CDK90]. This formula can be evaluated using its Mu-Calculus characterization.

The product is defined by its transition relation R, and set of initial states S_0. Let

1. $R = \lambda((s, s'), (r, r')). \exists \sigma. (\Delta(s, \sigma, r) \wedge \Delta'(s', \sigma, r'))$,
2. $S_0 = \lambda(s, s'). ((s = p_0) \wedge (s' = p'_0))$,

There is a sequence in the language of M but not in the language of M' if and only if there is an path of the product passing through B infinitely often, but not through B' infinitely often. Thus, $\mathcal{L}(M) \subseteq \mathcal{L}(M')$ iff

$$S_0 \Rightarrow AG\ A_{\{\lambda(s,s'). B(s)\}} F \lambda(s, s'). B'(s')$$

Another possible approach to the language containment problem makes use of the transitive closure of the transition relation. First, we remove from the product structure all transitions that begin or end with a state in B'. That is, let

$$T = \lambda((s, s'), (r, r')). [R((s, s'), (r, r')) \wedge \neg B'(s') \wedge \neg B'(r')]$$

The transitive closure of this relation is

$$T^+ = \mu Q. \lambda(x, y). [T(x, y) \vee \exists u. (Q(x, u) \wedge Q(u, y))]$$

This is the set of all pairs (x, y) of states of the product such that x can reach y without passing through B'. This holds for the pair (x, x) if and only if x is on a cycle not passing through B'. If there is any such x in B, and x is reachable, then there is a path passing through B but not B' infinitely often, hence there is a sequence in $\mathcal{L}(M)$, but not in $\mathcal{L}(M')$. The converse is also true. Hence, $\mathcal{L}(M) \subseteq \mathcal{L}(M')$ if and only if

$$\neg EF\lambda(s,s').\ (T^+((s,s'),(s,s')) \wedge B(s))$$

The EF operator in this formula can also be evaluated using the transitive closure, since

$$EFp = \lambda x.\ (p(x) \vee \exists y.\ (R^+(x,y) \wedge p(y)))$$

6.5 RELATED RESEARCH

The Mu-Calculus model checking algorithm gives us a fairly general framework for describing symbolic verification algorithms. Because of the excitement generated by Binary Decision Diagrams, there has been a fair amount of work in this general area (that is, verification of sequential systems using OBDD's). Most of the algorithms can be framed in the Mu-Calculus with the addition of a few special operators on Binary Decision Diagrams.

A variant on the symbolic model checking technique for CTL was proposed by Bose and Fisher [BF89b]. Their technique, which is limited to deterministic finite state machines, represents the transition relation of the machine by a vector of Boolean functions δ, and uses Bryant's *Compose* operation to compute $EXp = p(v_i \leftarrow \delta_i)$. They do not report experimental results using this technique for practical circuits. A similar technique was proposed by Coudert and Madre [CMB91].

Burch, Clarke and Long report on the use of early quantification (cf. section 3.4.2) for both disjunctive and conjunctive transition relations [BCL91b, BCL91a]. They use the term "partitioned transition relations" for this. The technique is somewhat limited in the case of conjunctive transition relations, because existential quantification only distributes over conjunction in the special case when one of the conjuncts does not depend on the variable being quantified. Nevertheless, there are cases where the support of the component relations is sufficiently disjoint to make this technique effective.

The basic technique is the following: assume we wish to compute $\exists v.\ \bigwedge_i f_i$, where $v = (v_1, \ldots, v_k)$ is a vector of variables and $f = (f_1, \ldots, f_m)$ is a vector of Boolean functions. Since conjunction is associative and commutative, we can combine these functions in any order we choose. In addition, if at any time there is a variable occurring in only one function, we can quantify that variable out, since $\exists w. (p \wedge q)$ is equivalent to $(\exists w.p) \wedge q$ when q does not depend on w. Since quantification tends to reduce OBDD size by reducing the number of variables, the strategy is to combine the functions in such an order that variables can be quantified out as soon as possible.

Burch Clarke and Long use a fixed order determined by the user for combining the functions. They show that this is quite effective for pipelined data path circuits, and an asynchronous stack circuit, improving the asymptotic performance as the circuit size increases. For the DME circuit, the asymptotic performance of this method was not as good as a method using a disjunctive transition relation, but it can be more efficient for small rings.[3] It was found most efficient to group the components of the transition relation and combine each group in advance, thus avoiding some computation at each step.

For disjunctive transition relations (interleaving models), Burch, Clarke and Long introduce a modified search order that tends to reduce the representation of the reached state set. In a breadth first search, the representation of this set is complicated by the fact that the after n steps, the number of steps taken by each process is constrained to sum to n. This produces an artificial correlation between the states of otherwise independent processes (cf. section 3.4.2). To counter this, one can modify the search order, searching first all of the states reachable by transitions of one subset of the system processes, then the next, and repeating this process until a fixed point is reached. This technique, called "modified breadth first search", was effective in reducing the OBDDs representing the reached state sets for an asynchronous stack circuit, but was found not to be as effective as the "conjunctive partitioning" method. For the DME circuit, the modified breadth first search method was faster up to about 16 cells, but had slower asymptotic performance. The grouping of processes into subsets was manual.

Coudert, Madre and Berthet have developed a variety of techniques for computing the reachable state set for the purpose of deciding the equivalence of two deterministic finite state machines [CBM89], in a system called PRIAM. Equivalence of two state machines is important information to know if one machine is intended to be an optimization of the other, or an implementation of the other using different gate primitives. In the method of Coudert, Madre and Berthet, a finite state machine is represented by a pair of vector Boolean functions. The function $\delta(v, w)$ yields the next state vector as a function of the

---

[3] Personal communication.

current state vector v and the input vector w. The function $\lambda(v, w)$ yields the output vector as a function of v and w. The equivalence of two state machines is tested by creating a combined machine in which both machines receive the same input vector. The output is a single bit which is true if and only if the output vectors of the two machines are equal. The reachable states of this combined machine are computed. If in all reachable states the output is true, the two machines are equivalent, since no input sequence can produce differing output sequences from the two machines.

The set of reached states is computed as the limit of an increasing series of approximations, starting with the initial state. The set of states reachable in one step from a set S is computed by a function called $Imag$, where $Imag(\delta, S) = \{s \mid \exists v, w : v \in S, \delta(v, w) = s\}$. Most of Coudert, Madre and Berthet's efforts are applied to computing the $Imag$ function without resort to representing the transition relation as an OBDD, which they claim is generally intractable. Their approach begins by reducing the problem of computing the image of a set to computing the range of a function. This is done using an OBDD operation called $Constrain$. The $Constrain$ operator takes two Boolean functions f and g, and returns a function $f' = Constrain(f, g)$ with the following property: for all x', $f'(x') = f(x)$, where x is the nearest Boolean vector to x' (according to a suitable distance metric) such that $g(x) = 1$. If we let $\delta' = Constrain(\delta, S)$, then the image of S via δ is just the range of δ'.

Coudert and Madre suggest two methods for computing the range of δ'. The first is called range partitioning. In this approach, we pick the lowest remaining variable in the ordering (call it v_i), and, and divide the problem into two subproblems, depending on the output of function δ'_i. Thus,

$$(Range(\delta'))(v_i \leftarrow 0) = Range(Constrain(\delta', \neg \delta'_i))$$
$$(Range(\delta'))(v_i \leftarrow 1) = Range(Constrain(\delta', \delta'_i))$$

Note that for any function f,

$$Constrain(f, f) = 1 \text{ and}$$
$$Constrain(f, \neg f) = 0$$

so each recursion effectively eliminates one component function of δ'. The recursion terminates when all of the components of δ' are constants.

The other approach, called domain partitioning, is to divide into subproblems based on the value of one of the inputs to δ'. Thus,

$$Range(\delta') = Range(\delta'(v_i \leftarrow 0)) \lor Range(\delta'(v_i \leftarrow 1))$$

Again, the recursion terminates when all of the components of δ' are constants.

Both of these strategies are special cases of a general strategy where one chooses a *cover*, which is a pair of functions h_1 and h_2 such that $h_1 \vee h_2 = 1$, and then computes the recursion

$$Range(\delta') = Range(Constrain(\delta', h_1)) \vee Range(Constrain(\delta', h_2))$$

In the case of range partitioning, $h_1 = \delta'_i$ and $h_2 = \neg \delta'_i$. In the case of domain partitioning, $h_1 = v_i$ and $h_2 = \neg v_i$. It is suggested that other covers may be useful as well. As with other OBDD techniques, a table of pairs $(\delta', Range(\delta'))$ is kept to avoid solving the same subproblem twice. This table is not as effective as the in the case of the standard OBDD operations, however, since the number of possible subproblems is exponential in the number of state variables. Coudert and Madre suggest several optimizations for increasing the hit rate in this table.

A further optimization introduced by Coudert and Madre is to use an OBDD function called *Restrict* to reduce the size of the reached state set before applying the *Imag* operator. The *Restrict* operator takes two functions f and g, and produces a function $f' = Restrict(f, g)$ such that for all x, if $g(x) = 1$, then $f'(x) = f(x)$, otherwise $f'(x)$ is arbitrary. Usually (but not always), the size of f' is less than the size of f. We note that if R_i is the set of states reachable after i steps of the machine, then

$$\begin{aligned} R_{i+1} &= R_i \vee Imag(\delta, R_i) \\ &= R_i \vee Imag(\delta, Restrict(R_i, \neg R_{i-1})) \end{aligned}$$

As a result, the size of the arguments of *Imag* can sometimes be reduced using *Restrict*.

Coudert and Madre report experimental results for computation of the set of reachable states for a variety of small sequential circuits (mostly ISCAS[4] sequential benchmark circuits). Computing the set of reached states can be useful for generating test patterns or "don't care" conditions for logic optimization [TSL+90]. Unfortunately, they do not use their techniques to actually test the equivalence of two state machines, so it is unknown whether the technique is useful for this purpose. They have not studied the asymptotic performance of their techniques for classes of circuits, so it is not possible to determine how the improvements gained by the various optimizations scale with the size of the state machine.

---
[4] International Symposium on Circuits and Systems

A related technique for computing the reachable states of a machine was introduced by Touati et al. [TSL+90]. They use a conjunction of component relations to represent the transition relation, along with early quantification. However, they combine this technique with the *Constrain* operation of Coudert et al. This reduces the problem of computing the image of a set via a relation to that of computing the codomain of a relation. A series of approximations A_i to the reachable states is computed, such that

$$A_{i+1} = A_i \vee \lambda y.\ \exists x.\ (\bigwedge_j Constrain(R_j, A_i))(x, y)$$

where R is a vector of component relations, each relation determining the new state of one state variable. Touati et al. find this technique to be superior to using the transition relation directly and to using the *Imag* operation of Coudert et al. for computing the reachable states of the benchmark circuits minmax and sbc, somewhat slower for key, and roughly the same for cpb.32.4. It would be interesting to know for the cases where an improvement was obtained, how much was due to the use of *Constrain* and how much to the use of early quantification. Touati et al. have also suggested partitioning complex next-state functions into the composition of a sequence of smaller functions. This could be useful for circuits containing multipliers, or other functions which have no compact OBDD representation.

Touati, Brayton and Kurshan also report a technique for testing language containment of ω-automata using OBDDs [TBK91]. They use the L-automaton model of Kurshan [Kur86], and an algorithm similar to the one described in section 6.4 that uses the transitive closure of the transition relation.

Another way that equivalance between two finite state machines can be established is by computing the equivalence relation on states (see chapter 8). Lin et al. describe OBDD based algorithms for computing the equivalence relation [LTN90] and how the equivalence relation can be used for computing "don't care" conditions for logic optimization. In a later paper [LN91], Lin shows how this relation (represented as an OBDD) can be used for state minimization, using an operator which takes an equivalence relation and returns a relation which maps every state onto the least element of its equivalence class.

Bryant and Seger have taken an an approach to formal verification using OBDDs based on symbolic simulation [Bry88, BBS90, BS90]. The symbolic simulator is similar to an ordinary logic simulator, except that the inputs are symbolic values (variables) rather than numeric values, and the outputs are given as symbolic functions in terms of these variables. These functions are represented by OBDDs. The simulation method gains a great deal in efficiency by using an abstract interpretation of the circuit model. This abstraction uses a lattice consisting of the three values 0, 1 and X, where X is the least upper bound of 0 and 1. The circuit operations such as AND and OR are abstracted in such a way as to be monotonic with respect to this lattice. Therefore, the output of the abstract simulation is always an upper bound on the output of the concrete simulation. In many cases, a large number of the inputs and initial values of state variables can be replaced by X without sacrificing the particular circuit property being proved. The art in this technique is to decompose the specification in such a way that each part can be verified using only a small number of symbolic values and X everywhere else. The simulation technique is limited to a logic with only next-time operators. These formulas can be verified using symbolic simulations of finite execution sequences. This rules out proving properties such as liveness, fairness or deadlock freedom, but allows safety properties to be proved using invariants.

Bose and Fisher have demonstrated a technique for using representation functions to verify sequential circuits using OBDDs [BF89a]. A representation function maps each state of the implementation to a state of the specification (which is also a circuit). Symbolic simulation techniques can be used to show a kind of single step equivalence between the implementation and specification *vis à vis* this relation. As in the method of Bryant and Seger, this proof can be decomposed into parts in such a way that each part requires only a small number of symbolic variables, with the remaining circuit nodes initialized to X. Typically, an invariant is also required, since the single step equivalence only holds over the reachable state space of the implementation. This technique is also limited in that it cannot prove liveness or deadlock properties.

Long and Grumberg have introduced an abstraction technique using OBDDs which is more general than simply introducing X values [CGL92]. Their technique uses an OBDD to express the relation between the abstract and concrete domains. The abstract transition relation is automatically derived using OBDD techniques from the concrete transition relation. This can be done in a compositional way to reduce the number of symbolic variables that are required. A variety of abstractions have been put to use in this way. For example, a binary number can be represented by its remainders modulo a set of relatively prime numbers. This has allowed the use of the Chinese remainder theorem to prove the correctness of a multiplier circuit. In another case, a single bit was used to represent whether a given binary number in a circuit is equal to a given symbolic binary value. In this way the entire function of the arithmetic unit was abstracted away, allowing a data pipeline circuit with 64 64-bit registers to be verified. This abstraction technique is quite general, and is closely related to more classical abstraction techniques [Kur87]. The difference is that function graph methods are used to actually compute the abstract transition relation, rather than giving this relation *a priori*.

7
INDUCTION AND MODEL CHECKING

This chapter deals with the verification of systems that are generated out of similar or identical finite state components, using a simple rule. Systems of this type are commonplace – they occur in bus protocols and network protocols, I/O channels, and many other structures that are designed to be extensible by adding similar components. In particular, we are interested in verifying properties that hold independent of the number of components in the system. We may for example have used a model checker to verify some property of a multiprocessor with a fixed number of processors. Would that property continue to hold if we added one more processor. One hundred? One thousand? Clearly, at some point we will be unable to answer the question with the model checker alone. It would be far preferable to settle the question once and for all, regardless of the number of processors.

Of course, the usual way to approach this problem would be to use induction on the number of processors. This requires an invariant (or inductive hypothesis). Such invariants are often complex and difficult to come by, however, and proving the invariant would require the use of a general purpose theorem prover, with all of the difficulties that this entails. On the other hand, if the invariant is itself a finite state process, we can use model checking techniques to prove the invariant. In addition, it may be that the invariant as a finite state process is not very different from our model of a processor, since n processors on a bus may not look very different from one (or two, or three). This idea can serve as a guide for constructing a first guess at the invariant. Counterexamples generated by the model checker can also provide clues to finding a correct invariant.

7.1 THE GENERAL FRAMEWORK

Induction over systems of processes can be put in a fairly general framework, which is independent of the mechanics of the process model, relying only on certain algebraic properties of the operators for combining processes. Let us assume that we have a collection of processes, and a collection of operators acting on processes. In a typical process model, we have some form of parallel composition operator, some form of operator for renaming signals, and perhaps a hiding operator, which makes a given signal invisible to the outside. The exact choice of operators is not material here, however. We require only that the operators be monotonic with respect to a reflexive transitive relation $\leq$ on processes. The idea of this order is that if $p \leq q$, then p is in some sense more specific, or has less possible behavior, than q. The properties we wish to verify should be preserved as we descend the order.

As an example of induction on processes, suppose we have a parallel composition operator $\|$ on processes, which is monotonic with respect to a pre-order $\leq$. In this case, we can apply the following induction rule:

$$\frac{p \leq q \qquad q \| p \leq q}{p \| \cdots \| p \leq q}$$

Think of the inequalities $p \leq q$ and $q \| p \leq q$ as substitution rules. If $p \leq q$, we can safely substitute p for any occurrence of q in a given term. This substitution will only make the term lesser in the pre-order. Thus, we can always substitute p for q on the lesser side of an inequality. For example, if $q \| p \leq q$, we have

$$\begin{aligned} q \| p &\leq q \\ (q \| p) \| p &\leq q \\ ((q \| p) \| p) \| p &\leq q \\ &\cdots \end{aligned}$$

If $p \leq q$, we can substitute p for q, giving us $p \| \cdots \| p \leq q$. We call q a *process invariant*. Other induction rules can be generated, based on other substitutions. For example, assume we have a parallel composition operator $\|$ and a renaming operator ϕ, both monotonic with respect to $\leq$. Then we have

$$\frac{\phi(q) \| r \| p \leq q}{\phi(\phi(\cdots) \| r \| p) \| r \| p \leq q}$$

Induction and model checking 131

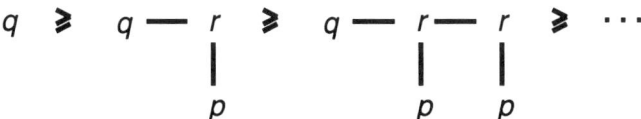

Figure 7.1 Processes generated by safe substition

Given a collection of substitution rules, we can inductively generate a class of processes from any process q. For example, figure 7.1 depicts the first few processes in the class generated by the above induction rule. Every process in the class is smaller than q in the pre-order. Thus, any properties of q which are preserved as we descend the pre-order are inherited by all the processes in the class. The key is to choose a pre-order that preserves the properties we are interested in verifying. The most straightforward way to do this is to choose a class of properties that we wish to preserve, and then define the pre-order accordingly.

Suppose we wish to preserve all properties expressible in the logic CTL. In this case, the pre-order we obtain is a degenerate one, which partitions the Kripke models into a set of incomparable equivalence classes. To see this, assume towards a contradiction that p satisfies every CTL formula satisfied by q, and there is some formula f satisfied by p but not by q. In this case, it follows that q satisfies $\neg f$. This implies, however, that p satisfies $\neg f$, a contradiction. Hence, if $p \leq q$, then p and q satisfy the same set of CTL formulas. Since CTL characterizes Kripke models up to bisimulation [BCG87], it follows that p and q are bisimular. This is unfortunate, since we do not want our induction framework to apply only to classes of Kripke structures that are observationally equivalent. In general, we would like to treat systems whose behavior becomes more specific as we add processes to the system.

One way to do this is to use a subset of the logic. For example, suppose we choose to preserve those formulas which use only universal path quantifiers. This subset is called ∀-CTL [GL91]. A formula in CTL is also in ∀-CTL if driving the negations in to the literals results in a formula without the E path quantifier. Examples of ∀-CTL formulas are

$$AG \neg EGp \equiv AGAF \neg p$$
$$\neg EGEXp \equiv AFAX \neg p$$

Examples of CTL formulas which are not in ∀-CTL are

$$AG \neg AFp \equiv AGEG \neg p$$

$$\neg EGAXp \equiv AFEX\neg p$$

Clearly, if a formula f contains path quantifiers, then f and $\neg f$ cannot both be in ∀-CTL. Grumberg and Long [GL91] have shown that if p satisfies every ∀-CTL formula satisfied by q, then q simulates p, and conversely. Simulation is easily shown to be reflexive and transitive. Thus simulation is a pre-order suitable for inductive proofs of ∀-CTL formulas. Let $p \leq q$ iff q simulates p. This gives us the following induction rule:

$$\frac{q \models f \quad (f \in \forall\text{-CTL}) \\ p \leq q \\ q \parallel p \leq q}{p \parallel \cdots \parallel p \models f}$$

as well as other rules engendered by various systems of safe substitutions. Recall from section 6.4 that simulation is the greatest relation between the states of q and the states of p such that if x simulates y, then:

1. x and y agree on the atomic propositions, and

2. every successor of y is simulated by a successor of x.

A Kripke model q simulates p if every initial state of p is simulated by some initial state of q. Since this relation can be expressed as a greatest fixed point in the Mu-Calculus (see section 6.4), it can be verified automatically using the symbolic model checking technique. The fact that simulation is not symmetric allows us more flexibility in constructing systems using substitution rules than we would have using bisimulation.

7.2 INDUCTION AND SYMBOLIC MODEL CHECKING

In our symbolic model checking framework, a process consists of a transition relation R, represented by a Boolean formula, and a set of initial states I, also represented by a Boolean formula. Now suppose that the only operators we allow for combining processes are logical conjunction ($R_1 \wedge R_2$) and the prefix operator ($a.b$) of section 4.3. We would like to use simulation as our pre-order,

Induction and model checking

so let $p_1 \leq_{AP} p_2$ iff

$$\forall x.(I_1(x) \Rightarrow \exists y.(I_2(y) \wedge S(x,y))) \tag{7.1}$$

where

$$S_0 = \lambda x,y. \wedge_{i \in AP} (x_i = y_i) \tag{7.2}$$

$$\begin{aligned} S = \nu Q.\lambda x,y.(S_0(x,y) \wedge \forall x'.R_1(x,x') \Rightarrow \\ \exists y'.(R_2(y,y') \wedge Q(x',y'))) \end{aligned} \tag{7.3}$$

and AP is some subset of the variables we want to use as atomic propositions in ∀-CTL specifications. If $p_1 \leq p_2$, then p_1 satisfies every formula of ∀-CTL(AP) satisfied by p_2. To complete our induction framework, we require that our process operators, conjunction and prefixing, be monotonic with respect to simulation. Unfortunately, this is not the case for conjunction. For example, suppose

$$\begin{aligned} AP &= \{a\} \\ p_1 &= 1 \\ p_2 &= (b \iff \neg b') \\ p_3 &= (a \iff b) \end{aligned}$$

In this case, p_1 and p_2 are indistinguishable with regard to the atomic propositions. In particular, $p_1 \leq_{AP} p_2$ and $p_2 \leq_{AP} p_1$. On the other hand, $(p_1 \wedge p_3) \leq (p_2 \wedge p_3)$ is false. Along all paths of $(p_2 \wedge p_3)$, the atomic proposition a is alternately true and false, whereas there are paths of process $(p_1 \wedge p_3)$ along which a is always true. Therefore the later satisfies $AF \neg a$, whereas the former does not.

In order to make the conjunction operator monotonic, we require a side condition. We can prove that

$$p_1 \leq_{AP} p_2 \Rightarrow (p_1 \wedge p_3) \leq_{AP'} (p_2 \wedge p_3) \tag{7.4}$$

provided $\sup(p_2) \cap \sup(p_3) \subseteq AP$ and $AP' \cap \sup(p_2) \subseteq AP$, where $\sup(p)$ is the set of variables occurring in p. In other words, conjunction is monotonic provided we are combining processes that communicate only via the atomic propositions (with respect to which p_2 simulates p_1). Our example above does not qualify, because b occurs in both p_2 and p_3 and is not an atomic proposition. The case $b \in AP'$ would also not be allowed.

Prefixing is also not monotonic, except under certain conditions. For example, suppose

$$AP = \{a.b\}$$

$$p_1 = 1$$
$$p_2 = b$$

Since $p_1 \leq_{AP} p_2$, we would like to have $a.p_1 \leq_{AP} a.p_2$, but this does not hold. This is equivalent to saying $1 \leq_{AP} a.b$, which is clearly false since $a.b$ satisfies $AGa.b$, but 1 does not. We can, however, easily prove the following:

$$p_1 \leq_{AP} p_2 \Rightarrow a.p_1 \leq_{a.AP} a.p_2 \qquad (7.5)$$

These two results allow us to construct an infinite series of successively lesser processes by a derivation like the following:

$$\begin{aligned} (a.q \wedge p) &\leq_{AP} & q \\ a.(a.q \wedge p) &\leq_{a.AP} & a.q \\ (a.(a.q \wedge p) \wedge p) &\leq_{AP} & (a.q \wedge p) \\ &\vdots & \end{aligned}$$

provided that $\sup(a.q) \cap \sup(p) \subseteq a.AP$ and $AP \cap \sup(a.p) \subseteq AP$, a condition that can easily be checked automatically. We can construct such a series in the SMV language, using the cache protocol as an example.

7.3 EXAMPLE: THE GIGAMAX PROTOCOL

We would like to formulate a safe substitution rule that generates an arbitrary number of Gigamax processors attached to a cluster bus. Thus, our process p in the above derivation will represent a single processor and its connection to the bus. The latter part computes the wired-OR of signals from the processor with signals from the remainder of the bus. Signals from the remainder of the bus will be prefixed with a. These will not be visible from the outside (*ie.*, they will not be atomic propositions in AP), but they will be accessible to p.

The invariant process q can be chosen in any way we like, with no restrictions. However, a good first guess would be to let q be a bus with one processor. Figure 7.2 illustrates the resulting substitution rule, generating an arbitrary nmumber of processors on a bus. This choice of q is like saying that, viewed from the outside, any number of processors can be simulated by one processor. This is unlikely to be true, but it may hold with slight modifications. If it

Induction and model checking 135

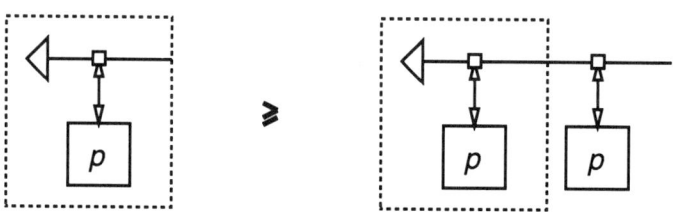

Figure 7.2 Substitution generating processors on bus

turns out not to be true, our strategy will be to make q more general by adding choices to its behavior, guided by the counterexample that the model checker produces.

The SMV code representing the processes p and q is listed in figure 7.3, along with the specification that tests the premise of our derivation, $(a.q \wedge p) \leq_{AP} q$.

Essentially, we are testing whether one processor can mimic the actions of two processors as seen from the bus. Checking this produces a counterexample in which one of the two processors reaches the owned state, then the second processor issues a read command. This behavior cannot be produced by a single processor. To fix this problem, we replace the processor model p with a new model r which is modified so that a processor is allowed to issue a read command in the owned state. It then sets its own "snoop" flag, and enters the shared state on a read-shared, and the owned state on a read-owned. Testing this new invariant produces another counterexample in which the first processor reaches the owned state, then issues a read command (thus setting its snoop and waiting bits), then the second processor issues a read command. One processor alone cannot produce this behavior, since it cannot issue a second read command while its waiting flag is set. We modify the processor model r to s, which is allowed to have this behavior. Note that this is behavior is safe *vis-a-vis* cache coherence, since the second read command is blocked by the waiting flag which is already set. With this modification, we have a correct invariant.

Using this invariant, we can check properties of the system in $\forall$-CTL, using the invariant q in place of a bus with any number of processors. The substitution rule can be applied as many times as necessary to produce a system with an arbitrary number of processors while preserving all of the verified properties. We can also refer these properties back to our original model p by showing that the generalized processor model s simulates p. That is, we can test $p \leq_{AP} s$.

```
MODULE p

VAR
  p : processor(CMD, REPLY-OWNED, REPLY-WAITING, REPLY-STALL);

--arbitration
ASSIGN p.master := !master union 0;

--pass signals on to rest of bus
DEFINE
  a.cmd :=
    case
      master : cmd;
      1 : p.cmd;
    esac;
  a.master := master | p.master;
  a.reply-owned := reply-owned | p.reply-owned;
  ...

--get accumulated wire-OR signals from rest of bus
DEFINE
  CMD := a.CMD;
  REPLY-OWNED := a.REPLY-OWNED;
  ...

MODULE q

ISA p;

-- end of bus -- return accumulated wire-OR signals
DEFINE
  CMD := a.cmd;
  REPLY-OWNED := a.reply-owned;
  ...

MODULE main

SPEC
  q simulates (a.q & p) on {master,cmd,CMD,...}
```

Figure 7.3 Substitution rule for adding one processor

Induction and model checking 137

If this holds, substituting p for every occurence of s will preserve any $\forall$-CTL property.

It should be observed that one property of the protocol verified in chapter 5 is not in $\forall$-CTL. The absense of deadlock formula contains an existential path quantifier. To prove this property using induction, it would be necessary to use bisimulation equivalence instead of the simulation pre-order. This would allow us to create substitutions which preserve all CTL properties. Unfortunately, it would not be possible to use the strategy of generalizing the original model p until an invariant is reached, since the generalized model s would not be bisumulation equivalent to the original model.

7.3.1 Computing simulation relations

There are a few techniques that can improve the efficiency of this process. The simplest is to note that simulation between two states implies that they agree on the values of the atomic propositions. Therefore, in practice, there is no need to use two copies of the atomic propositions. Instead we can use the following calculation: $p_1 \leq_{AP} p_2$ iff

$$\forall AP, x.(I_1(AP, x) \Rightarrow \exists y.(I_2(AP, y) \wedge S(AP, x, y))) \tag{7.6}$$

where

$$\begin{aligned} S = \nu Q.\lambda AP, x, y.(\forall AP', x'.R_1(AP, x, AP', x') \Rightarrow \\ \exists y'.(R_2(AP, y, AP', y') \wedge Q(AP', x', y'))) \end{aligned} \tag{7.7}$$

and x, x', y, y' contain copies of only the variables which are not in AP. This reduces the number of variables used to represent the simulation relation.

As in the case of CTL model checking, a forward reachability calculation can be useful as a restriction on the simulation relation – if a pair of states is not reachable, then whether these states simulate each other or not has no efect on simulation between initial states (as one can easily prove by induction). There are at least two possible ways using reachability. One is to compute the reachable states for p_1 and p_2 independently. Thus,

$$RS_i = \mu Q.\lambda x.(I_i(x) \vee \exists y.(Q(y) \wedge R_i(y, x))) \tag{7.8}$$

$$S_0 = \lambda AP, x, y.(RS_1(AP, x) \wedge RS_2(AP, y)) \tag{7.9}$$

$$\begin{aligned} S = \nu Q.\lambda AP, x, y.(S_0(AP, x, y) \wedge \forall AP', x'.R_1(AP, x, AP', x') \Rightarrow \\ \exists y'.(R_2(AP, y, AP', y') \wedge Q(AP', x', y'))) \end{aligned} \tag{7.10}$$

On the other hand, we can take into account the fact that all pairs of states that are simular agree on the atomic propositions. Thus, we can restrict the simulation relation to those pairs of states which are reachable in parallel, along paths agreeing on the atomic propositons. In other words, let

$$\begin{aligned} I &= \lambda AP, x, y.(I_1(AP, x) \wedge I_2(AP, y)) \\ R &= \lambda AP, x, y, AP', x', y'.(R_1(AP', x', AP, x) \wedge R_2(AP', y', AP, y)) \end{aligned}$$

and

$$S_0 = \mu Q \lambda AP, x, y.(I(AP, x, y) \vee \\ \exists AP', x', y'.(Q(AP', x', y') \wedge R(AP, x, y, AP', x', y')))$$
(7.11)

Either way, the same result is obtained for simulation between initial states. The different versions of the algorithm may have different efficiencies, however.

There is another way of testing simulation which is conservative, in the sense that false negatives are possible, but not false positives. We observe that if two states x and y are n-simular, but not $(n+1)$-simular, there is a pair of paths π_x and π_y, starting with x and y respectively, such that for $0 \leq i \leq n$, $\pi_x(i)$ and $\pi_y(i)$ are 1-simular, but $\pi_x(n+1)$ and $\pi_y(n+1)$ are not. This means, for one thing, that we can use π_x and π_y as a counterexample if the simulation is false. It also means that if no such π_x and π_y exist, then the simulation holds. This gives us a conservative check that is akin to testing string language containment. To wit, $p_1 \leq_{AP} p_2$ if (but not only if) $RS \Rightarrow S_1$, where

$$S_1 = \lambda AP, x, y. \forall AP', x'.(R_1(AP, x, AP', x') \Rightarrow \\ \exists y'.R_2(AP, y, AP', y'))$$
(7.12)

In fact, if p_2 is deterministic (ie., if $R(x, y) \wedge R(x, z) \wedge S_0(y, z) \Rightarrow S(y, z)$), then the "only if" is true as well. In this case, we are testing both string language containment and simulation – the two are equivalent [GL91].

7.3.2 Compositionality in SMV

The reader may recall from the semantics of SMV that when a process is not running, any variable whose next value is assigned in that process keeps the same value (unless the assignment is conditional). This may have seemed like an unusual way to define processes. In other languages (UNITY, for example [CM88], and many concurrent procedural languages) it is possible for any variable of a process to be changed by another process. This would have very unfortunate consequences for induction over processes and other compositional

Induction and model checking 139

methods, however. Using these methods, we wish to reason about a process in the absense of other processes in the system. If the internal variables of a process cannot be guaranteed by local considerations to remain constant while other processes are running, then we can say almost nothing about the process by reasoning about it in isolation. In particular, the construction of an invariant process would be impossible, since the simulation relations would have to hold even if the processes were initialized to random states by an outside process. In effect, it would be necessary to put conditions on the behavior of outside processes which would be equivalent to what is implicit in SMV the SMV semantics, and then to prove separately that the outside processes actually conform to these conditions.

7.4 INDUCTION IN OTHER MODELS

We can set up an induction framework for a variety of models by establishing a pre-order and a set of monotonic process operators. For models of concurrent automata (such as the s/r model [Kur85]), the natural pre-order is language containment.

In the s/r model, a process is an automaton which accepts infinite strings over a Boolean algebra. There are two natural operators over this class of processes. The automaton product operation simulates parallel execution, while Boolean algebra homomorphisms can be used to induce a renaming or abstraction of the variables by which processes communicate. Kurshan shows that both of these operations respect the relation of language containment between automata [Kur86]. An example of induction in this framework can be found in [KM89].

Induction can also be applied in process algebras like CCS [Mil80] which are based on two-way synchronization. In this case, there is a variety of plausible process relations, including observational equivalence, weak observational equivalence, and a number of pre-order relations on processes. An induction example using the "may" pre-order for CCS processes can also be found in [KM89].

7.5 RELATED RESEARCH

A number of methods have been proposed in the past for extending automatic verification to systems composed of an arbitrary number of similar or identical processes.

The first to approach this question were Browne, Clarke and Grumberg [BCG86], who extended the logic CTL to a logic called *indexed CTL*. This logic allows the restricted use of process quantifiers as in the formula $\bigvee_i f(i)$, which means that the formula f holds for some process i. Restricting the use of these quantifiers and eliminating the next-time operator makes it impossible to write a formula which can distinguish the number of processes in a system. By establishing an appropriate equivalence between a system with n processes and a system with $n + 1$ processes, one can guarantee that all systems satisfy the same set of formulas in the indexed logic. This method was used to establish the correctness of a mutual exclusion algorithm by exhibiting a bisimulation relation between an n-process system and a 2-processes system, and applying model checking to the 2-process system.

A disadvantage of the indexed CTL method is that the bisimulation relation must be proved in an *ad hoc* manner. Finite state methods cannot be used to check it because it is a relation between a finite-state process and a process with an arbitrary number of states. Clarke and Grumberg dealt with the problem of establishing a bisimulation by introducing the notion of a *process closure* P^*. This process must be derived by hand, and have the property that $M_r \parallel P^*$ is equivalent to $M_{r+1} \parallel P^*$ for some small r. This can be verified mechanically. Shtadler and Grumberg took this notion a step further by introducing *network grammars* to describe classes of finite state systems. This technique used an indexed form of linear temporal logic, and required that the processes on the left and right hand sides of each grammar rule be equivalent in an appropriate sense.

The requirement that all systems generated by the grammar be equivalent seems to be a rather strict limitation, however. The method of this chapter, which uses a pre-order rather than an equivalence, was first proposed by Kurshan and McMillan [KM89], and simultaneously by Wolper and Lovinfosse [WL89]. Around the same time, Burch was also applying a similar idea to Dill's trace theory for speed-independent circuits.[1]

Another method for proving properties of systems of identical processes is due

---

[1] Personal communication

to German and Sistla [GS]. It uses a linear-time temporal logic for specifications (again, the next-time operator is not allowed) and is fully automatic. By means of a distinguished "control" process, it is possible to check some global properties (although process quantifiers are not present in the logic). Unfortunately, because the decision algorithm is doubly exponential in the process size, this method has not been applied in practice.

A system called GORMEL has been created by Marelly and Grumberg, implementing the techniques of [SG89]. GORMEL uses context free grammars to describe systems of processes. This is fairly similar to the use of module substututution rules in SMV. There are a number of differences between the systems, however. GORMEL is oriented towards verification of distributed algorithms. It uses a model of transition systems with pairwise synchronized actions, as in CCS. This model is not well suited for describing digital systems – first because most signals in hardware are broadcast to more than one location, and second because many signals are exchanged back and forth between components of a system in a single clock cycle. The difficulty of reducing this two way exchange of many signals to a single atomic action would make it extremely cumbersome to create a CCS-like model for a system like the Gigamax.

Another difference is in the logic – GORMEL uses an indexed version of LTL without next-time called LTL^2. As in indexed CTL, it is not possible to nest process quantifiers. Because of the ability to use process quantifiers, it is possible to express some properties which are not expressible in CTL, for example that if a proposition p is true in some process, then it is eventually true in all processes.

For the process relation, GORMEL uses a form of stuttering equivalence rather than simulation. This places fairly strong requirements on the allowable grammar rules. In particular, it is not possible in such a system to take the approach taken here of successively generalizing a component process in order to obtain an invariant, since the required relation between the left and right hand sides of the grammar rule is a symmetric one. The GORMEL approach will work if the various systems generated by the grammar can be distinguished only by stuttering (arbitrary repetition of the same state labeling).

A final difference between the systems is, of course, that SMV is based on symbolic model checking methods. This is not clearly an advantage, however, since the state explosion problem may not be very severe for the small number of processes that tend to be involved in induction rules.

8

EQUIVALENCE COMPUTATIONS

In this section, we consider the problem of computing a symbolic representation of the equivalence relation between the states of two finite state machines, or between states of the same machine. In the former case, the relation can be used to determine the equivalence of the two machines, while in the latter case, as Lin *et al.* have observed [LTN90], the self equivalence relation can be used in optimizing the logic or register usage of the machine.

8.1 STATE EQUIVALENCE

We use a standard notion of the equivalence of states of finite Mealy machines. Two states are equivalent if and only if for all input sequences, they yield the same output sequence. The following is an alternate characterization: equivalence is the greatest relation between states such that if x is equivalent to y, then for all inputs, the output in state x is equal to the output in state y, and the successor state of x is equivalent to the successor state of y. Let $\delta(x, z)$ be the function which determines the next state, as a function of current state x and current input z, and let $\gamma(x, z)$, be the function that determines the current output. In the Mu-Calculus, the equivalence relation R_ω is

$$R_\omega = \nu R.\ \lambda x, y.\ \forall z.\ (\gamma(x, z) = \gamma(y, z) \land R(\delta(x, z), \delta(y, z))) \qquad (8.1)$$

Using the standard fixed point approach, we can evaluate this relation by a sequence of approximations $R_0, R_1, \ldots$, where R_i is the set of state pairs which are equivalent for all input sequences of length i. This sequence is characterized by the recurrence

$$R_1 = \lambda x, y.\ \forall z.\ (\gamma(x, z) = \gamma(y, z)) \qquad (8.2)$$

and
$$R_{i+1} = \lambda x, y. \ \forall z. \ (R_i(x,y) \wedge R_i(\delta(x,z), \delta(y,z))) \qquad (8.3)$$
This is simply the standard $O(n^2)$ algorithm for computing state equivalence of Mealy machines. The problem of determining whether two Mealy machines are equivalent in their initial states can be approached in two ways – either their equivalence relation can be computed, or the state space of their product can be exhausted by a forward search. The number of iterations required for the former approach can be substantially less, however. In the trivial case of an n-bit counter, the number of iterations in the forward search is is exponential in n, while one step suffices to reach a fixed point in the equivalence calculation, since all states are distinguished by their outputs.

It is immediately seen that the crucial step in calculation 8.3 is the substitution of vector functions $\delta(x,z)$ and $\delta(y,z)$ into R_i. The most obvious way to accomplish this is to use Bryant's *Compose* algorithm. Some other possible methods are introduced in this section. Computing the OBDD representation for a composition of functions is an NP-hard problem (cf., section 3.3), thus we expect no good general solutions to the problem. Another tractability issue is whether the approximations to the equivalence relation can be compactly represented using OBDDs. There is no guarantee of this, of course, but there is some reason to believe, *a priori*, that it may often be the case. First of all, for single-machine equivalence, if all distinct states are distinguishable, then the equivalence relation is the identity relation, which can be represented by a linear size OBDD, provided the component variables of x and y are interleaved in the variable ordering. It also seems plausible that the equivalence relation will often be simply a logical conjunction of independent relations, each corresponding to some modular component of the system. In this case, the OBDD representation will also be compact, provided the variable ordering conforms to the modular structure of the machine. In any case, we will see examples of fairly complex machines whose equivalence relations are expressed compactly in OBDD form.

8.1.1 Algorithm using restrictions

Because of the basic difficulty of computing compositions of OBDDs, it is useful to have some restrictions on the result in order to be able to solve the problem. Fortunately, the decreasing nature of the series of approximations defined in 8.3 provides a constraint on the result of the substitution, since each approximation

Equivalence computations

R_{i+1} is strictly contained in R_i. We can use this fact by rewriting 8.3 as

$$R_{i+1} = \lambda x, y. \ \forall z. \ (R_i(x,y) \wedge (R_i(\delta(x,z), \delta(y,z)) \downarrow R_i)) \qquad (8.4)$$

where $\downarrow$ represents the *Restrict* operator introduced by Coudert, Madre and Berthet [CBM89]. This operation produces a function which agrees with the first argument $R_i(\delta(x,z), \delta(y,z))$ when the second argument R_i is true, attempting to minimize the OBDD size. The restriction can be used to varying advantage, depending on the algorithm used for substitution.

8.1.2 Iterative abstraction algorithm

Another way to provide a restriction on the equivalence relation is first to find the equivalence relation of an abstracted machine. We choose the abstraction in such a way that the equivalence relation of the abstract machine is strictly weaker than the equivalence relation of the original machine. Thus, we can compute the equivalence relation of the abstract machine first, and use it as a restriction in computing the equivalence relation of the original machine. In particular, we can abstract the machine by choosing a subset V of the state variables, and at each approximation, quantifying out the remaining variables existentially. That is, let V^c be the complement of V, and let

$$R_1^V = \lambda x, y. \ \exists V^c. \ \forall z. \ (\gamma(x,z) = \gamma(y,z)) \qquad (8.5)$$

and

$$R_{i+1}^V = \lambda x, y. \ \exists V^c. \ \forall z. \ (R_i^V(x,y) \wedge (R_i^V(\delta(x,z), \delta(y,z)) \downarrow R_i^V)) \qquad (8.6)$$

It is trivial to see that each approximation in the series R^V is strictly weaker than the corresponding approximation in R. It follows that R_ω^V, the greatest fixed point, is weaker than R_ω. Therefore, we can restrict the entire calculation of R_ω to only those state pairs satisfying R_ω^V. In addition, we can use a series of subsets $V_1 \subset V_2 \subset \cdots \subset V_k$ where V_k is the set of all state variables, restricting the first approximation in each series R^{V_i} to the equivalence relation for the previous subset. Thus, we let

$$R_1^{V_j} = R_\omega^{V_{j-1}} \wedge \lambda x, y. \ \exists V_j^c. \ \forall z. \ (\gamma(x,z) = \gamma(y,z)) \qquad (8.7)$$

and

$$R_{i+1}^{V_j} = \lambda x, y. \ \exists V_j^c. \ \forall z. \ (R_i^{V_j}(x,y) \wedge (R_i^{V_j}(\delta(x,z), \delta(y,z)) \downarrow R_i^{V_j})) \qquad (8.8)$$

We will refer to this as the iterative abstraction algorithm for computing the equivalence relation. By adding only a few variables to each successive subset, we can in some cases obtain fairly strong restriction, which allows the substitution to be computed more efficiently. In other cases the equivalence relation obtained for the abstracted machine may be trivial, since abstracting out the variables in V^c may result in all machine states appearing equivalent at the outputs. This is especially likely if the abstracted variables hold important control information that enables machine registers to be observed at the outputs. Nonetheless, we can show cases where this incremental approach is greatly more efficient than the basic algorithm.

8.2 METHODS FOR FUNCTIONAL COMPOSITION

This section considers methods for substituting functions for variables in OBDDs. This operation is referred to by Bryant as *Compose*. It is the syntactic mechanism corresponding to functional composition. As such, it has a number of applications apart from finding the equivalence relation of finite state machines, including the evaluation of CTL formulas [BF89b]. Most of the algorithms presented here for this purpose have been modified to take as an extra argument a restriction on the result, in the hope that efficiency can be obtained by combining these two operations. We consider the problem of calculating $f(g_1,\ldots,g_n) \downarrow R$, where f, $g_1,\ldots,g_n$ and R are all Boolean functions.

8.2.1 "bottom-up" substitution

This is the method originally proposed by Bryant for his *Compose* algorithm, but with a restriction on the result. In this method, we view each OBDD node in f as a gate, which computes the function "if v_i then h else l" or equivalently, $(\neg v_i \wedge l) \vee (v_i \wedge h)$. Having substituted the functions $g_1,\ldots,g_n$ for the variables in l and h, we can then compute the result for f using the standard $\vee$ and $\wedge$ operators. The basic bottom-up substitution algorithm is shown in figure 8.1.

Note that the restriction operator is used at each step to simplify the result. Since each subproblem is solved only once, the number of recursive calls to bottom-up is $|f|$.

```
function bottom-up(f,R)
    if f is a leaf then return f
    if bottom-up(f,R) has already been solved then return old solution
    else [f is a triple (v_i, f_l, f_h)]
        l = bottom-up(f_l, R)
        h = bottom-up(f_h, R)
        return ((¬g_i ∧ l) ∨ (g_i ∧ h)) ↓ R
end
```

Figure 8.1 Bottom-up substitution algorithm

8.2.2 Domain partitioning

The domain partitioning strategy is so named because it divides the problem into two subproblems by partitioning the domain of the functions $g_1, \ldots, g_n$ according to the value of one of the variables. The operation proceeds in several steps.

First, we observe that if any of $g_1, \ldots, g_n$ are constants, we can immediately substitute these values into f, since substitution by a constant is a linear time operation which can only reduce the size of the OBDD. We use the fact that if $g_i = c$, where c is 0 or 1, then

$$f(g_1, \ldots, g_n) = f(v_i \leftarrow c)(g_1, \ldots, g_n) \tag{8.9}$$

Next, we observe that we can eliminate any argument position on which f does not depend, thus obtaining a smaller problem with the same result. We can determine the set of variables on which f depends in linear time, since f depends on v_i if and only if v_i appears in some node in f.

If at this point the function f has been reduced to a constant, we are done. Otherwise, we split the problem into two cases and recurse. We choose the first variable v_i occurring in $g_1, \ldots, g_n$, and apply Shannon's expansion, obtaining two subproblems

$$\begin{aligned} l &= f(g_1(v_i \leftarrow 0), \ldots, g_n(v_i \leftarrow 0)) \\ h &= f(g_1(v_i \leftarrow 1), \ldots, g_n(v_i \leftarrow 1)) \end{aligned}$$

As in other OBDD algorithms, the result is an OBDD $r = (v_i, l, h)$, provided $l \neq h$, otherwise $r = l = h$. Needless to say, we use a hash table, caching the results of subproblems so that the same subproblem is not solved twice. With caching, the complexity of the algorithm is $O(|f| \times \prod |g_i|)$.

Making use of the restriction R in this algorithm is straightforward. If $R = 0$, the result can be any function at all, so we simply return 0. Each time we partition the problem into subproblems, we also split R into two cases, $R(v_i \leftarrow 0)$ and $R(v_i \leftarrow 1)$. The restriction has the effect of cutting off the recursion each time a 0 leaf is reached in R.

8.2.3 Sequential substitution

This is perhaps the simplest approach to substitution; it transforms a simultaneous substitution problem into a sequence of substitutions. This is done by replacing each variable v_i in the OBDD for f by a new variable v'_i. Having done this, it is safe to perform the substitutions of each function g_i for v'_i in any order, since none of the functions $g_1, \ldots, g_n$ depends on any variable being substituted. Substitution of a function for a single variable can be accomplished as follows:

$$f(v'_i \leftarrow g_i) = \exists v'_i . [(v'_i \iff g_i) \land f] \qquad (8.10)$$

This approach can also make effective use of a restriction. The restriction operator operator may in fact be applied after each substitution step if desired, potentially reducing the size of the intermediate results. In the case of the iterative abstraction algorithm, the fact that some of the variables in the result will later be quantified out existentially can also be put to use. We can move the existential quantifiers for these variables inside the conjunction, thus quantifying the abstracted variables out of the term $(v'_i \iff g_i)$ before applying the conjunction. This may weaken our result somewhat, since $[\exists x. a] \land [\exists x. b]$ is weaker than $\exists x.[a \land b]$, but it can produce significant reductions in the size of the intermediate results. The final result of the equivalence algorithm is unchanged, since it is computed with no variables abstracted.

8.3 EXPERIMENTAL RESULTS

This section presents the results of applying the various equivalence relation algorithms to several example state machines, with a range of complexity. The results are compared to published results for the same circuits by Touati *et al.*

Equivalence computations 149

machine	mtd	result (nodes)	b-u (secs)	d-p (secs)	seq (secs)	Touati (secs)	Lin (secs)
sbc	iter	361	2054	> 10K	2415	2903	
cpb32	iter	95	45.5	14.4	54.4	14.1	12.10
key	iter	167	342	1738	884	5706	175.20
minmax10	dir	89	197	255	204		
minmax20	dir	59	3.0	4.5	6.7		
minmax30	dir	89	6.2	9.0	15.7		

Table 8.1 Equivalence calculation times

(computing only the reachable states) and Lin et al. (computing the equivalence relation). In all cases, it is self-equivalence that is calculated. It would be interesting to have some results in this section on calculating the state equivalence relation between two different implementations of a given machine, but unfortunately, such examples were lacking. The three different approaches to OBDD substitution are compared, for each example. Where possible, the direct algorithm is used, otherwise, the iterative abstraction algorithm is used. For example, the state equivalence relation for the machine sbc was calculated using iteration, but could not be calculated directly. Table 8.1 gives the total execution time in seconds, while table 8.2 gives the total number of OBDD nodes used.[1] The latter numbers are not very reliable, since they depend to some extent on arbitrary choices about when to scavenge unused cells and cache entries. However, if the available memory limit of 190,000 nodes is exceeded, it is certain that all of the nodes in use were necessary for the computation, since all available nodes were scavenged when the memory limit was reached. The columns give the following information: the name of the circuit, the method used (direct or iterative), the size of the equivalence relation, and the time or space needed for each of the three substitution methods (bottom-up, domain partitioning, and sequential). The times are for a LISP implementation running on a 1 MIP minicomputer. The final two columns give the results obtained by Touati et al. and Lin et al. for the same circuit. These times are for C language implementations running on a DEC 5400 and IBM R6000 respectively.

It would seem that for the circuits cpb, key and minmax, which have regular structures with no control registers, there is no clear choice as to which substitution algorithm is best. The bottom-up algorithm tends to provide the best

---
[1] Actually, function graphs with negated arcs were used for this calculation [Bil87], hence the number of nodes may be slightly smaller than what would be obtained using OBDDs.

machine	mtd	result (nodes)	b-u (nodes)	d-p (nodes)	seq (nodes)
sbc	iter	361	22184		34609
cpb32	iter	95	8202	4225	11295
key	iter	167	13328	24868	11563
minmax10	dir	89	16589	17815	17566
minmax20	dir	59	8538	8492	9190
minmax30	dir	89	11952	11886	10001

Table 8.2 Equivalence calculation space

performance with the least memory usage, but there are a number of exceptions. The machine sbc, which is somewhat more complex, is a more interesting case. Here bottom-up and sequential both provide fairly efficient solutions, although the iterative method was required in both cases to solve the problem. The domain partitioning approach fails to terminate after 10,000 seconds. In the first stage of the iterative algorithm, domain partitioning produced over 100,000 subproblems for a final result of approximately 100 nodes. Obviously, many different subproblems with identical results are being solved. The difficulty is that there is no easy way to identify equivalent problems. It is worth mentioning the the limit on the size of the cache for this method was 5000 entries. With an unbounded cache, the performance of the algorithm may be much better (a matter of theoretical interest at best, since an unbounded cache cannot be provided). It should also be noted that the results for minmax are somewhat anomalous, since the 10-bit version seems to be substantially more complex than the 20- and 30-bit cases. This is explained by the fact that the output functions of these different versions were not the same. In the 20- and 30- bit versions, the outputs appear to depend only on the "last" register, and not the "min" and "max" registers. It is also interesting to observe that for minmax10, not all of the states are distinguishable, that is, the equivalence relation is not the identity.

Comparing these results to those of Touati et al., it is interesting to note that the self-equivalence relation can be computed in less time than the reachable states for sbc and key (taking into account the difference in machines speeds of roughly a factor 10, the equivalence method seems to be about one order of magnitude faster for sbc, and two orders of magnitude faster for key). Of course, the information obtained by the two methods is not the same. It seems, however, that in some cases where the set of reachable states is not obtainable,

Equivalence computations 151

the equivalence computation may still provide useful information for logic optimization. The results of Lin *et al.* seem to be roughly comparable for the machines key and cpb32 (again, taking into account the difference in machine speeds). It is not clear from the Lin *et al.* article which substitution method was used, since two were mentioned. The one benchmark for which the iterative method was required to produce a result was sbc, but unfortunately Lin *et al.* do not report a figure for this machine. Also, because of the fact the the 20- and 30-bit versions of minmax had modified output functions, it is not possible to compare figures for this benchmark. As a result of these ambiguities, it difficult to draw conclusions about the effectiveness of the iterative abstraction method, except to say that in one case (sbc) it was the only method that successfully computed the equivalence relation.

9
A PARTIAL ORDER APPROACH

This chapter deals with the question of how independence of events in concurrent systems contributes to the state explosion problem, and how a partially ordered representation can avoid this problem.

Consider, for example, the following simple game: the playing board consists of a single row of spaces; the game begins with a number of playing pieces on space X, and the object is to move all of the pieces to space Y, by moving each piece one space at a time. This is hardly a game at all, since the playing pieces can be moved from X to Y in any order, without affecting the outcome. However, consider writing a computer program to play this game. The standard approach would be to use a depth first search – a program that enumerates all possible combinations with one move, then two moves, then three, *etc.*, until a solution is found. This approach leads rapidly to a deep quagmire, since the number of combinations of this simple game is astronomical (that is, exponential in the number of playing pieces). One might imagine improving the depth first strategy by keeping a list of all visited board configurations to avoid visiting them again, but this stratagem is also hopeless.

In fact, our simple game is quite analogous to the problem of analyzing the behavior of concurrent finite state systems, as might arise in distributed protocols, parallel computations, or asynchronous digital circuits. Such systems are generally made of a number of relatively independent processes (*ie.*, the playing pieces) that synchronize or communicate only occasionally (such as in the final state of our game). Any attempt to analyze such systems by exhaustively enumerating the behaviors or reachable states must be limited to very small systems, because of the intractable combinatorics. However, in our simple game, ignoring the (irrelevant) ordering of moves leads to a much simpler (in fact trivial) analysis. Might such be the case with concurrent systems?

This question has been approached before [Val89, Val90, God90, GW91, YTK91]. Valmari and Godefroid have (independently) observed that the notion of *persistence* is an important one in determining when move ordering can be ignored. A persistent move is one which remains possible (is not blocked) regardless of any other moves which might be taken in the future. In our simple game, all moves are persistent, since we allow any number of playing pieces on a given space. Now, suppose that at a given point in our search, a move cannot be blocked, and cannot block any other move we might want to make in the future. Then we may as well make that move now as later, and we have no need to consider other branches of the search tree – all lead to the same possible outcomes.[1] The notion of persistence can be generalized to persistent sets of moves, in the sense that one move of the set must eventually be taken. In our simple game, since no move can result in the blocking of any other move, a search in the style Valmari or Godefroid can lead directly to the solution, without any branching in the search.

So far, so good, but life is not always so simple. In some cases, a straightforward analysis of the rules of play may determine that certain moves are persistent. In other cases, this information may be more difficult to obtain. For example, let us modify our simple game so that no space between X and Y may contain more than three pieces (for example, each piece might represent a process, which at each step requires a resource, of which only three are available). It is still true that the game is won regardless of the order in which legal moves are made. Further, all moves are persistent (and non-blocking) in the weak sense that if we keep playing the game, *eventually* any given piece will be able to move. Unfortunately, obtaining the latter information seems to be as difficult as the analysis of the game itself. In fact, this property is not an artificial one. In this chapter, we will be chiefly concerned with analyzing the behavior of asynchronous digital circuits. In our models of these circuits, the moves correspond to signal transitions; the nature of these rules is such that no transition or small set of transitions is inherently persistent – we must refer to the global behavior of the circuit to make this determination. Since this appears to be difficult, let us take a step back, and consider what other forms of analysis, besides a search tree, might be suitable to the problem.

At each step in constructing a search tree, we are forced to choose a move to take *next*. In other words, we are constructing *totally ordered* sequences of

---

[1] Actually, if the game contains cycles, such an approach might lead us to become stuck interminably in a cycle, missing a possible solution. The solution to this problem can be found in [God90]

moves. If this choice may result in an increase in information regarding the outcome (*ie.*, reduce the set of possible outcomes), we are forced to create a bifurcation in the tree, since we wish to consider all possible outcomes. This in turn leads to an increase in the number of combinations or states that we must search. We should note, however, that this choice of a total ordering is mandated by the structure of the search tree, and not by the rules of the game itself. These latter may allow us to choose a partial rather than a total order on moves, and thereby avoid unnecessary bifurcation in the search. Such an analysis requires a richer structure than a search tree − one that represents both a partial order on moves or transitions, and choice, or bifurcation in the search. One such structure, called a *pomtree*, has been invented by Probst and Li [PL89, PL90, PL91], and has been used in the verification of asynchronous circuits.[2]

Another framework for a partially ordered analysis has been provided by Nielsen, Plotkin and Winskel [NPW81]. They describe the unfolding of a model called a Petri net into an infinite net of a restricted form called an occurrence net. From the occurrence net, one can derive a partial order (one might say a causal order) on events, and a conflict relation between events which corresponds to bifurcation or choice. Taken together, these two relations form a prime event structure. Nielsen, Plotkin and Winskel describe the relation between event structures and domains of information as defined by Scott. Here, however, we are concerned with more mundane matters. Petri nets, while somewhat restricted in their expressiveness, can be used to model the behavior of asynchronous circuits in such a way as to capture the causal relation between signal transitions. We will discuss an algorithm that unfolds a sufficient fragment of the corresponding occurrence net to reveal important properties of the circuit, such as potential hazards and deadlocks, if such exist. We will also see that the unfolding approach can in some cases avoid the exponential explosion of states that results from the total ordering of transitions in standard search algorithms.

9.1 UNFOLDING

Let us begin with an informal introduction to Petri nets and unfolding. A Petri net is a game board of sorts. The spaces on the board are drawn as circles and called *places*. These may hold any number of game pieces, which are drawn as dots and called *tokens*. The rules for moving the tokens are given by a set of *transitions*, which are drawn as lines or boxes. An arrow from a place to a

---
[2] A pomtree is a tree whose edges are labeled with partially ordered multisets of events.

transition indicates that the transition removes a token from the place, while an arrow in the opposite direction indicates that the transition places a new token on that place. Removal occurs before placement, so a transition cannot be used until and unless there is one token at the tail of each arrow entering the transition. Later, we will see how this token game can be used to model an asynchronous circuit, using two places to model each wire. A token on one represents a logical zero on the wire, while a token on the other represents a logical one, and transitions moving tokens from one to the other correspond to rising and falling transitions on the wire.

Unfolding a Petri net is a game that can easily be played on a piece of paper, though it quickly becomes tedious. First, place some initial tokens on your Petri net. For each of these tokens, make a copy of the place on which it resides in the occurrence net (label the copies appropriately). Now, repeat the following process *ad infinitum*:

1. Choose a transition from your Petri net and call it t.

2. The set of places in your Petri net which have arrows going to t is called its *preset*. For each place in the preset of t, find a copy in the occurrence net and mark it with a token (if you can't find a copy, go back to step 1). Do not chose the same subset twice for a given t.

3. Two places x and y in the occurrence net are said to be *concurrent* if

    - You cannot follow any path of arrows from x to y or *vice versa*.
    - There is no third place z from which you can reach both x and y, exiting z by different arrows (this is called *conflict*).

    If any of the places you marked are *not* concurrent, go to step 1.

4. Make a copy of t in the occurrence net. Call it t'. Draw an arrow from every place you marked in the occurrence net to t'. Erase the tokens.

5. The set of places in your Petri net which are at the head of arrows originating at t is called its *postset*. For each place in the postset of t, make a copy in the occurrence net (label it, of course) and draw an arrow from t' to it.

If you make your choices fairly and have infinite time, you will build the unfolding of your Petri net (see figure 9.1). In the process, you will probably notice that in purposes, as soon as you choose a copy of a given place in the

A partial order approach

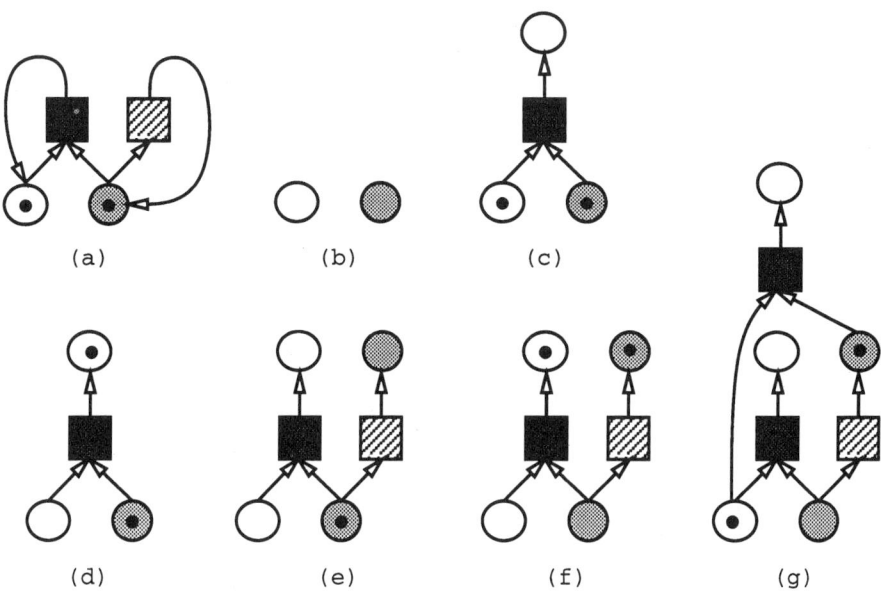

Figure 9.1 Unfolding a net: (a) the original net, with marking, (b) starting point of the occurrence net, (c) after one complete iteration, (d) chosen places are not concurrent, (e) after a second complete iteration, (f) chosen places are in conflict, (g) after a third complete iteration.

preset of t, it helps to black out (temporarily) any places in the occurrence net which are not concurrent with your choice. In this way you rapidly narrow down the choices, and avoid having to go back to the beginning in step 3. To see the advantage of this algorithm over an exhaustive state space search, the reader might want to make a Petri net model of the simple game from the introduction, and produce the unfolding (hint: use a special transition for the end game that removes all the tokens from Y – multiple arrows from a place to a transition are allowed).

There are several facts that are worth observing about the unfolding. The first is that the unfolding is an acyclic graph, defining a partial order on its nodes. We might think of this as providing a causal order on events (copies of the transitions). Second, bifurcations occur naturally in the structure where actual choice occurs (as opposed to the superfluous choice of the total order of unrelated transitions). This is the notion of conflict alluded to in step 3. Because of it, we are not put in the position of determining *a priori* which transitions are impersistent, and hence where bifurcations must occur. A final unfortunate fact, however, is that the unfolding may be infinite. If we wish to use it to infer properties of the original net, such as whether a given place may eventually receive a token or not, we will have to make do with a finite fragment. Such a finite fragment can be constructed, but we will require a certain amount of formalism to prove the adequacy of this fragment for our puroposes. In particular, we need to precisely define the relationship between the unfolding and the behaviors of the net.

We begin with the standard definition of a Petri net. It simplifies matters if we assume that there can be only one arrow from a given place to a given transition, though this assumption is not necessary.

Definition 4 *A (Petri) net is a triple $N = (P, T, F)$, where*

1. *P is a set of places,*
2. *T is a set of transitions (disjoint from P),*
3. *$F \subseteq (P \times T) \cup (T \times P)$.*

The preset of transition t, denoted $^\bullet t$ is the set of places p such that $p \; F \; t$. The postset of transition t, denoted $t^\bullet$ is the set of places p such that $t \; F \; p$. The notation $t_1^\bullet t_2$ means that $t_1^\bullet$ and $^\bullet t_2$ are incident.

An arrangement of tokens on a Petri net (a state of the token game) is usually called a *marking*. Since there can be any number of tokens on a given place, and tokens are indistinguishable, we will think of a marking mathematically as a multiset of places, or a map from the set of places to the natural numbers.

Definition 5 *A marking of a net (P, T, F) is a multiset on P. A marked net has an associated initial marking I.*

We now define recursively the legal sequences of moves in the token game. The postset of a sequence of transitions (with respect to the initial marking) is the marking that remains when the sequence of transitions is fired. One property of a net that we will be interested in is the set of reachable markings. A marking is reachable if it is the postset of some legal sequence of transitions. In particular, we would like to infer from the unfolding certain properties of the reachable markings of the original net.

Definition 6 *The firing sequences of a marked net N, and their postsets, are defined recursively as follows:*

1. *ϵ (the null sequence) is a firing sequence, and $\epsilon^\bullet = I$ (the initial marking).*

2. *if π is a firing sequence, σ is a transition, and $(\pi^\bullet) \supseteq (^\bullet\sigma)$, then $\pi\sigma$ is a firing sequence, and $\pi\sigma^\bullet = (\pi^\bullet) - (^\bullet\sigma) + (\sigma^\bullet)$.*

A marking M is reachable exactly when there is a firing sequence π such that $M = \pi^\bullet$.

When we unfold a Petri net, we build a structure which will be called a labeled occurrence net. This is a Petri net in which every place and transition is labeled with a corresponding place or transition in the original net. The occurrence net is a specialized form of net which must satisfy certain restrictions. First, it must be well founded, meaning that the arrows cannot be followed backward infinitely from any point. Second, it must have no forward conflict, meaning that two arrows may not converge on the same place. Third, no event may be in conflict with itself, and fourth, no two events with the same label may have the same preset. The reader might want to verify that the informal unfolding procedure above in fact constructs an occurrence net.

Definition 7 *A (P, T) labeled occurrence net N' is a Petri net (P', T', F'), with a labeling function L' which maps P' onto a set P and T' onto a set T. The net must have the following properties:*

1. *Well foundedness: every subset of T' must contain a minimal element with respect to F'^*.*

2. *No forward conflict: for all places $p \in P'$, $p \in t_1^\bullet$ and $p \in t_2^\bullet$ implies $t_1 = t_2$.*

3. *No self-conflict: for all $t_1, t_2, t_3 \in T'$, $t_1 F'^* t_3$ and $t_2 F'^* t_3$ and $^\bullet t_1 \cap ^\bullet t_2 \neq \emptyset$ implies $t_1 = t_2$.*

4. *No redundancy: for all $t_1, t_2 \in T'$, $L'(t_1) = L'(t_2)$ and $^\bullet t_1 = ^\bullet t_2$ implies $t_1 = t_2$.*

The occurrence net derived by unfolding a given Petri can be characterized as follows:

Definition 8 *If N is a marked net, then the unfolding of N is the maximal (P, T) labeled occurrence net (up to isomorphism) satisfying the following:*

1. *for all $t' \in T$, $L'({}^\bullet t') = {}^\bullet L'(t')$ and $L'(t'{}^\bullet) = L'(t')^\bullet$, and*

2. $L'(P' - \mathrm{cod}(F')) = I$.

We will not concern ourselves here with proving that the procedure described above actually builds this net. It will be of interest, however, to prove some results concerning the relation of unfoldings and firing sequences. To do this, we introduce the notion of a *configuration*. A *configuration* of an occurrence net is a set of events representing a possible partially ordered run of the net. There is a requirement of causality for configurations: if we start with an event that is in the configuration, and trace a path backward along the arrows, all events we encounter must also be in the configuration. We call this requirement backward closure. A configuration must also be free of conflict: two events in the configuration may not be at the heads of arrows originating at the same place.

Definition 9 *Let N' be a labeled occurrence net. A subset S of T' is a configuration of N' exactly when*

1. *It is backward closed: if $t_1 {}^\bullet t_2$, then $t_2 \in S$ implies $t_1 \in S$.*

2. *It is conflict free: for all distinct $t_1 \in S$, $t_2 \in S$, ${}^\bullet t_1$ and ${}^\bullet t_2$ are disjoint.*

Clearly, a configuration induces a partially ordered multiset on transitions:

Definition 10 *If S is a configuration of labeled occurrence net N', let*

$$\mathcal{P}(S) = (S, F'^* \cap S^2, L/S).$$

The structure $\mathcal{P}(S)$ is a pomset on transitions because of the well-foundedness of F'^*. As such it represents not a single firing sequence, but an equivalence class of sorts – those firing sequences which are linearizations of $\mathcal{P}(S)$. It is less obvious, but also true, that a single firing sequence can correspond to more than one configuration of the unfolding if the net is not safe, *ie.*, if it allows

A partial order approach

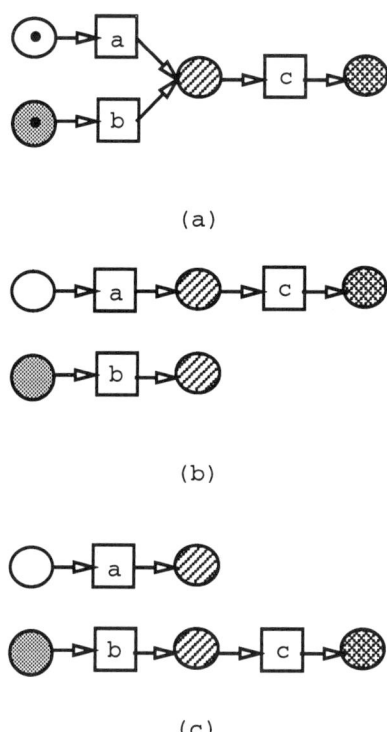

Figure 9.2 (a) a Petri net, (b) and (c) two configurations corresponding to the firing sequence *abc*.

more than one token to appear on a given place (see figure 9.2). This matter complicates somewhat the relation between configurations of the unfolding and firing sequences of the original net, but in the author's opinion, not sufficiently to justify restricting our consideration to safe nets.

The notion of configuration is evidently both more and less specific than firing sequence, for unsafe nets. However, there is a useful notion of equivalence we can draw between the two, with regard to postsets. We note that regardless of the linearization of the events in a configuration, the same number of tokens are removed from and placed on each place by the firing sequence. Thus a configuration has a well defined final state. This final state is determined by the postset of the configuration: those places on the frontier between events in the configuration and events not in the configuration. The multiset of their labels is a marking on the original net, which we call the *final state* of the configuration.

Definition 11 *Let S be a configuration of N'. The postset of S is the set of all places $p \in P'$ such that*

1. *for all $t \in S$, $p \notin {}^\bullet t$*
2. *for all $t \in T' - S$, $p \notin t^\bullet$*

The final state of S, denoted $\mathcal{F}(S)$, is $L'(S^\bullet)$ (the multiset of labels appearing in $S^\bullet$).

Our first theorem will show that the final states of all the configurations are exactly the reachable markings of the original net.

Lemma 4 *Given a labeled occurrence net N', if C and $C' = C \cup \{t'\}$ are configurations of N', then $C'^\bullet = C^\bullet - {}^\bullet t' + t'^\bullet$.*

Proof. Backward closure (definition 9) implies that ${}^\bullet t' \subseteq C^\bullet$. Lemma follows from definition 11.

Lemma 5 *Given a labeled occurrence net N', if C and $C' = C \cup \{t'\}$ are configurations of N', then $\mathcal{F}(C') = \mathcal{F}(C) - {}^\bullet L(t') + L(t')^\bullet$.*

Proof. From definition 8 (part 1), definition of $\mathcal{F}$ and previous lemma.

Lemma 6 *Given N', the unfolding of a net N, s a firing sequence of N, $t \in T$ such that st is a firing sequence of N, and C a configuration of N' such that $\mathcal{F}(C) = s^\bullet$, there exists $t' \in T'$ such that $C' = C \cup \{t'\}$ is a configuration of N', and $\mathcal{F}(C') = st^\bullet$.*

Proof. Suppose no such t' exists. Construct a new net N'' by adding event t', satisfyin definition 8 part 1, such that ${}^\bullet(t') \subseteq C^\bullet$. Since C is conflict free, t' is non-self-conflicting. Also, t' is non-redundant, since by the above lemma, any transition redundant with t' must prove this lemma, and we have assumed no such t' exists. Therefore, N'' is an occurrence net, satisfying points 1 and 2 of definition 8, thus contradicting the maximality of N'.

A partial order approach

Lemma 7 *Given N', the unfolding of a net N, s a firing sequence of N, C a configuration of N' such that $\mathcal{F}(C) = s^\bullet$, and $t' \in T'$ an event such that $C' = C \cup \{t'\}$ is a configuration of N', it follows that $sL(t')$ is a firing sequence, and $\mathcal{F}(C') = (sL(t))^\bullet$.*

Proof. From lemma 5 and definition 6.

Theorem 6 *Let N' be the unfolding of N. M is a reachable marking of N if and only if M is the final state of some finite configuration of N'.*

Proof. By induction on the length of firing sequences. The basis case derives from the firing sequence ϵ and the configuration $\emptyset$, using part 2 of definition 8. The induction step for the forward direction derives from lemma 6 and in the reverse direction from lemma 7.

9.2 TRUNCATED UNFOLDINGS

Having established that the configurations of the infinite unfolding represent exactly the set of reachable markings of the original net, we are now in a position to consider whether a finite fragment of the unfolding might be constructed which is sufficient to represent all of the reachable markings. On the surface, this is a reasonable proposition, since a bounded net[3] has a finite number of reachable markings. A construction yielding such a finite fragment could be used to answer such questions as whether a given set of places in the Petri net is coverable, meaning a state can be reached with at least one token on every place in the set. This information can be used to verify that an asynchronous circuit is hazard free. The same fragment of the unfolding can also be used to determine if the token game can ever reach a deadlock – a state in which no transitions are enabled, as we will see later.

The algorithm is based on the notion of a *local configuration* – any configuration which has a unique maximal element. The backward closure of any event t' (*ie.*, $F'^{*-1}(t') \cap T'$) is a configuration (owing to non-self-conflict of events) and is called the local configuration of t'.

---
[3] a net which generates no more than a fixed number of tokens on each place, for any firing sequence

Definition 12 *Let N' be a labeled occurrence net, and let $t' \in T'$. The local configuration of t', denoted $\lceil t' \rceil$ is the least backward closed subset of T', with respect to F', containing t'.*

A local configuration has, of course, a final state. We will call an event a *cutoff point* if there exists another event whose local configuration is smaller, but has the same final state. The intuition behind this definition is that any configuration containing the first event must be equivalent to a smaller configuration containing the second. Any configuration containing a cutoff point therefore adds no new reachable markings to the unfolding, thus a cutoff point may be excluded from the unfolding. This argument will be examined in detail shortly, but first let us modify the unfolding procedure, by adding one more step to the loop, between steps 3 and 4:

3.5. If there is any event t'' in the occurrence net such that $L'(\lceil t'' \rceil^\bullet) = L'(\lceil t' \rceil^\bullet)$ and $|\lceil t'' \rceil| = |\lceil t' \rceil|$, go to step 1.

Clearly, the test in this step can be done more quickly using a hash table to store the final states of the existing local configurations in association with their size. The algorithm terminates when we run out of choices in steps 1 and 2. This algorithm always terminates for bounded nets, and at termination, the configurations of the occurrence net represent exactly the reachable markings of the original net by their final states. The argument for this proposition runs roughly as follows:

Definition 13 *Let N' be an unfolding of a Petri net N. A transition $t' \in T'$ is a cutoff point of N' exactly when there exists $t'' \in T'$ such that*

1. $\mathcal{F}\lceil t' \rceil = \mathcal{F}\lceil t'' \rceil$, and
2. $|\lceil t'' \rceil| < |\lceil t' \rceil|$

The truncation of N' is the greatest backward closed subnet of N' containing no cutoff points.

Lemma 8 *Let N' be an unfolding of N, and let C be a configuration of N'. There exists a configuration S of N' such that C and S have the same final state, and S does not contain a cutoff point.*

A partial order approach

Proof. If C contains no cutoff point, lemma is trivial. Else, let t be a cutoff point contained in C. By definition, there is a transition t' in C such that $\lceil t \rceil$ and $\lceil t' \rceil$ have the same final state and $|\lceil t' \rceil| < |\lceil t \rceil|$. We can show by induction on $|C| - |\lceil t \rceil|$ that there exists a configuration C' containing t' such that the final states of C and C' are equal, and $|C| - |\lceil t \rceil| = |C'| - |\lceil t' \rceil|$. This implies that $|C'| < |C|$. We can therefore apply transfinite induction, by assuming the theorem holds with C' for C. This implies the existence of a configuration S not containing a cutoff point, having the same final state as C', and hence the same final state as C. □

Theorem 7 *Let N' be the unfolding of N. M is a reachable marking of N if and only if M is the final state of some finite configuration of the truncation of N'.*

Proof. Forward direction. By theorem 6, if M is reachable, then there exists C configuration of N' such that $\mathcal{F}(C) = M$. By the above lemma, there exists configuration S such that $\mathcal{F}(C) = M$, containing no cutoff points. The truncation must contain S, since otherwise there would be a greater subnet of N' containing no cutoff points. Reverse direction follows from theorem 6.

Theorem 8 *Let N' be an unfolding of N. If N is bounded, then the truncation of N' is finite.*

Proof. Given a transition t in the truncation, let π be a longest chain of events ending in t such $t_1 {}^\bullet t_2$ holds for each successive pair (t_1, t_2) in the chain. For each such pair, $|\lceil t_1 \rceil| < |\lceil t_2 \rceil|$. The length of π cannot be greater than the number of reachable markings of N, since otherwise the local configurations of two elements of the chain would have the same final state, hence the latter of the two would be a cutoff point. We can show by induction that the number of transitions for which π has a given length is finite, since if transition t has length n all the predecessors of t have length $n - 1$ or fewer, leaving us a finite number of combinations for the preset of t. Therefore the number of transitions in the truncation is finite. □

As an example of termination of the algorithm, consider the net of figure 9.3, which represents the dining philosophers. In this scenario, there are n concurrent processes (philosophers), each of which must acquire the use of two shared resources (forks) in order to execute its critical section (eating spaghetti). The

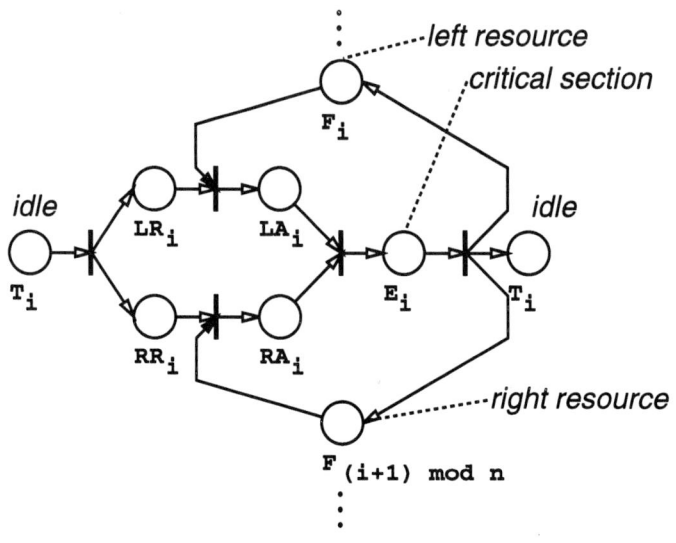

Figure 9.3 Dining philosophers net.

processes are organized in a ring, with each neighboring pair sharing one resource. Figure 9.4 shows the completed unfolding for the case of three philosophers ($n = 3$). The cutoff points are marked with an X. The local configuration of each of these transitions is equivalent to the empty configuration (initial marking). The size of the unfolding is not only bounded, but is linear in the number of philosophers. The number of states (reachable markings) is exponential however, as shown in table 9.1.

One of the questions we can easily answer using the truncated unfolding is

n	unfolding size (transitions)	reachable states
2	9	22
3	13	100
4	17	466
5	21	2164

Table 9.1 Unfolding size and number of states for Dining Philosophers

A partial order approach

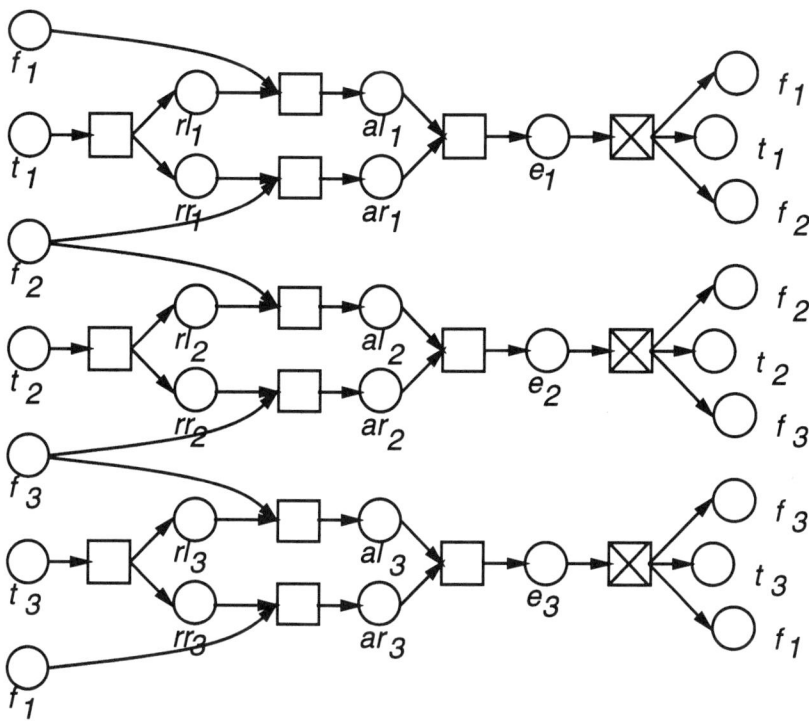

Figure 9.4 Unfolding of the dining philosophers net.

whether a given place is *coverable*, meaning that at least one token can be put on that place by means of some firing sequence.

Corollary 1 *Let N' be an unfolding of N. A place p in N is coverable iff there is some place p' in the truncation of N' labeled with p.*

Proof. By theorem 7.

Since we can test coverability of a single place in the net, we can easily test it for a subset of places by simply adding a transition to the original net whose preset is the set of places in question, and whose postset is a single place not otherwise used in the net. Coverability of this latter place is equivalent to coverability of the set.

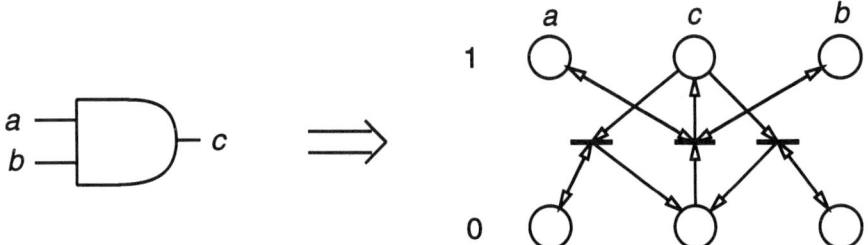

Figure 9.5 Translation from circuit to net

9.3 APPLICATION EXAMPLE

We now consider a more realistic example than the dining philosophers – a speed-independent [Sei80b] circuit designed to implement a distributed mutual exclusion (DME) protocol. The circuit was designed by Alain Martin [Mar85] and has been analyzed using an abstracted trace theoretic model by Dill [Dil88].

Networks of logic gates in speed-independent circuits are readily modeled by Petri nets. A network of n gates can be modeled by a Petri net of $O(n)$ places. When we model a network of gates as a Petri net, we introduce two places for each input of each gate. One represents the input in a logic low state, while the other represents the input in a logic high state. Transitions in the Petri net correspond to rising or falling transitions of gate outputs. A rising transition of a gate output removes all the logic low tokens from the inputs to which it is connected, and places tokens on the corresponding logic high places.

As an example, figure 9.5 shows a net fragment representing an AND gate with one fanout. When both inputs of the gate are at the logic high state, we can move a token from the place representing logic low at the output to the place representing logic high. Similarly, if either input is at the logic low state, we can move a token from the place representing logic high at the output to the place representing logic low.

It is important in designing asynchronous circuits to know that their behavior is free of dynamic hazards. A dynamic hazard occurs, for example, if the AND gate's output is enabled to rise while one of its inputs is enabled to fall. Such a situation can produce a pulse on the gate output which is not a legitimate digital signal, and thus violates the assumptions under which the gate is modeled as a digital device. The problem of whether or not this dynamic hazard can occur can be posed as a coverability problem, of that set of places enabling both transitions. Alternatively, since dynamic hazards correspond to dynamic conflicts in the unfolding, the problem can be solved by constructing the unfolding and examining it for dynamic conflicts, *ie.*, two transitions which are in conflict, and which may be simultaneously enabled. The DME circuit also uses special two-way mutual exclusion elements as components, which are immune to certain hazards. In checking the DME ring for hazards, we ignore conflicts between rising transitions of a mutual exclusion element's acknowledge outputs.

Figure 9.7 shows the results of the occurrence net unfolding procedure for the Petri net model of the DME circuit, for rings with one to nine cells. The depth of the occurrence net unfolding for the case of 5 cells was 141 transitions. The number of transitions in the unfolding, shown in part (a) of the figure, increases quadratically in the number of cells. This is because as the number of cells in the ring increases, a request must be relayed through a greater number of stages in order to obtain the token, in the worst case. At the same time, the number of cells which may request also increases. The occurrence net therefore grows in both width and depth in proportion to the number of cells. The time to construct the unfolding (running a LISP implementation on a Sun3 workstation) appears to increase quartically, as shown in part (b) of the figure. Finally, as we increase the number of cells in the ring, the number of reachable global markings increases exponentially, as shown in part (c) of the figure (on a logarithmic scale).[4] The number of states increases asymptotically by slightly less than a factor ten for each added cell. The fact that the truncated unfolding increases quadratically in size implies that this exponential increase in states is due only to the arbitrary ordering of independent transitions.

This is an interesting result, because it highlights a distinction between analysis using unfolding, and methods of Valmari [Val89, Val90], Godefroid [God90, GW91] and Yoneda [YTK91], which perform state space search using the notion of persistence to limit the search. Recall that a persistent transition is one which remains enabled regardless of the firing of other transitions. In the case of gates, any output transition that is enabled may become disabled by a transition on one of the inputs. In general, determining whether such a disabling transition can occur before the output transition occurs is a difficult problem, for which one must consider the global structure of the circuit. Experiments by Holger Schlingloff[5] have confirmed that the method of Yoneda provides little or no improvement over an ordinary search strategy.

---

[4] The number of reachable states was established using the symbolic model checking technique [BCM+90].
[5] Personal communication

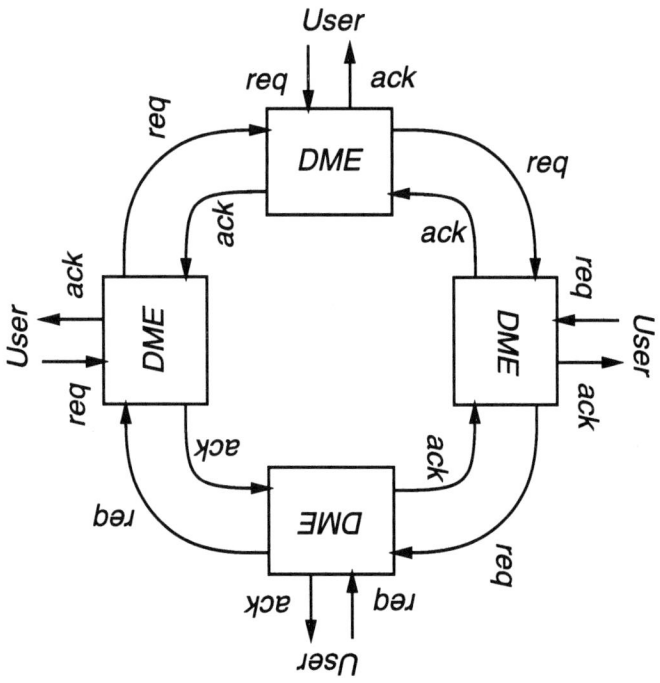

Figure 9.6 Distributed mutual exclusion circuit

There are, however, other methods of analysis which can avoid large state space searches, some of which do not directly exploit partial ordering of events. For example, the trace theory approach of Dill [Dil88] has been applied to the DME circuit. This requires an abstract model of the arbiter cell to be created by hand. The abstraction reduces the state explosion problem, but does not entirely solve it, since even with the reduced model, the number of states increases exponentially with the number of components. Probst [PL91] reports a method which requires quadratic space and time in the number of cells, though it requires a pomtree model of the circuit to be constructed by hand. Finally, the symbolic model checking technique [BCM+90] has also been applied to the DME circuit. The basic symbolic model checking algorithm requires cubic time and linear space (in the number of cells). Burch and Long[6] have obtained $O(n^{2.5})$ time for the DME using symbolic model checking with a modified search order [BCL]. This method requires some hand optimization,

---

[6] Personal communication

however. In any event, it appears that the symbolic model checking method yields somewhat better asymptotic performance for the DME circuit, though both methods effectively solve the state explosion problem. The unfolding method has an advantage over the symbolic model checking method in that no variable ordering or other heuristic information is required. It is not difficult to construct a variation on the dining philosophers for which there is no good variable ordering for symbolic model checking, but for which the unfolding is still linear space (in the number of philosophers). However, the author is presently unaware of any practical circuits for which this is the case.

9.4 DEADLOCK AND OCCURRENCE NETS

Besides coverability, another interesting problem for Petri nets is the question of deadlock. A *terminal marking* of a Petri net is one in which no transitions are enabled. Reachability of a terminal (or deadlocked) state cannot be framed in terms of the coverability problem. However, since the unfolding represents all reachable markings, a net has a reachable terminal marking if and only if its unfolding has a reachable terminal marking. The problem of existence of a terminal marking in an occurrence net is NP-complete. This is easily shown by reduction from 3-SAT.[7] To see this, consider the formula $(x_1+y_1+z_1)(x_2+y_2+z_2)\cdots(x_n+y_n+z_n)$ where each x_i, y_i and z_i is a positive or negative literal. Assume the formula has m variables. Let the positive literals be $l_1,\ldots,l_m$, and the negative literals be $\bar{l}_1,\ldots,\bar{l}_m$. In polynomial time, we can construct a net which has a terminal marking if and only if the formula is satisfiable. The initial marking of the net is a set of places $\{v_1,\ldots,v_m\}$. There is a place representing each positive literal $l_1,\ldots,l_m$ and each negative literal $\bar{l}_1,\ldots,\bar{l}_m$. For each variable v_i, there is a transition from v_i to l_i and from v_i to $\bar{l}_i$. For each conjunct $(x_i+y_i+z_i)$, there is a transition c_i, whose preset is $\{\bar{x}_i,\bar{y}_i,\bar{z}_i\}$. In other words, the transition c_i is enabled to fire if and only if $(x_i+y_i+z_i)$ is false. Thus, some transition c_i is enabled to fire if and only if the whole formula is false. The postset of each transition c_i is the single place $\{q\}$, and there is a transition from $\{q\}$ to $\{q\}$. Thus, if any c_i fires, the net may never reach a terminal marking. As a result, there is a terminal marking of the net if and only if the formula is satisfiable. For example, figure 9.8 shows the net constructed for the formula $(a+b+\bar{c})(b+c+\bar{d})$.

---
[7]Satisfiability of a Boolean formula in conjunctive normal form, with three literals in each conjunct.

The reader may easily verify that the size of the unfolding of such a net (up to the cutoff points) is linear in the size of the original net. In fact, it is essentially the same net, except the the place q occurs n times in the unfolding. Since all reachable markings of the original net occur as configurations of the unfolding, the unfolding has a terminal marking if and only if the formula is satisfiable. Hence 3-SAT is P-time reducible to reachability of a terminal marking of an unfolding. Since the configuration representing the terminal marking can be guessed in P-time in the size of the unfolding, and also tested in P-time, it follows that the problem is in NP, and hence NP-complete.

Interestingly, however, the problem is readily solved in practice even for very large unfoldings, using an algorithm based on techniques of constraint satisfaction search. The key observation which leads to this algorithm is that there is no terminal marking exactly when every configuration of the unfolding is a subset of some configuration containing a cutoff point. This is simply because if there is no terminal marking, then every configuration can be extended to a configuration which is arbitrarily large. A configuration C' can be extended to a configuration containing transition t' if and only if the union of C' and the local configuration of t' is a configuration. If it is not, then no set containing C' and t' is a configuration. If the union is not a configuration, we will say that C' and t' are in conflict. Hence, there is a terminal marking if and only if there is a configuration which is in conflict with every cutoff point. The search for such a configuration can be carried out using branch and bound techniques. For example, if a configuration C' is in conflict with a cutoff point t', there must be a transition $t'_1 \in C'$ which is in conflict with t'. Such a transition t'_1 will be called a *spoiler* of t'.

There exists a configuration in conflict with all of the all of the cutoff points (equivalently, there exists a terminal marking) if and only if there exists a configuration containing a spoiler for every cutoff point. The set of spoilers contained in this configuration will be called T_s. The following algorithm uses branch and bound techniques to find such a set T_s if one exists.

```
1  let B be the set of the cutoff points, T_s = ∅
2  while B is not empty do
3     let t the the element of B with the fewest spoilers
4     if t has no spoilers, then backtrack
5     choose an element t' from the spoilers of t
6     add t' to T_s
7     delete all transitions in conflict with T_s
8  end do
```

A partial order approach

Note that in line 3 of the procedure, the cutoff point with the smallest number of spoilers is chosen so that the number of choices in line 5 is minimized. Whenever a spoiler for a given cutoff point is chosen to belong to T_s in line 5, everything in conflict with T_s is eliminated from future consideration in line 7. Note that the cutoff points in conflict with T_s are also eliminated, which cuts down on the amount of future branching. Whenever there is a cutoff point with no remaining spoilers, the procedure backtracks, from line 4 to the most recent occurrence of line 5 where there are remaining choices. If there are no remaining choices, the procedure fails. When backtracking occurs, the the net is returned to the state it was in at the point where execution is being resumed. This backtracking is easily implemented by keeping a stack of the remaining choices for t' in each iteration of the loop, and marking each transition in the net with the level of the stack at the time it was "removed". Interestingly, if the procedure terminates successfully, the remaining net has the property that every maximal configuration corresponds to a class of terminal firing sequences. This makes it straightforward to extract such a sequence, by building a maximal configuration, and choosing any linearization of that configuration.

Because of the backtracking, this procedure is exponential (as it must be, if $\mathcal{P} \neq \mathcal{NP}$). However, this is only the worst case. The dining philosophers serve as an example of a case in which the exponential complexity is avoided. In fact, the procedure finds the terminal marking in time which is *linear* in the number of philosophers. This is easily seen by examining the unfolding of the Dining Philosophers net in figure 9.4. There is one cutoff point in this net for each process. Initially, each of these transitions has two spoilers, which correspond to the two resources required to enter the critical region being granted to the two neighboring processes. Regardless of which cutoff point is used first, the symmetry is then broken as the part of the net in conflict with one of the two spoilers is removed. This removes, in particular, the transition which granted one of the resources to the first philosopher, hence one of its neighbors now has only one spoiler, so there is only one choice available the next time line 5 is reached. After this spoiler is added to T_s, the remaining neighbor of the second philosopher now has only one spoiler. This process continues without backtracking until it has come full circle and the terminal marking is found. Note that if the cutoff point with the fewest spoilers were not chosen in line 3, the procedure might have examined an exponential number of candidates for T_s before a valid one was found.

For the DME circuit example, we find that the run time of the deadlock algorithm is 218 seconds for a ring of five cells, and 6600 seconds for a ring of 9 cells. Hence, even though the the algorithm is exponential in the worst case, in this case it runs in reasonable time for an unfolding of over 5000 transitions. It is clear that the branch and bound technique quickly narrows down the number of choices for this example.

9.5 CONCLUSION

We have observed that much of the complexity in the analysis of concurrent finite state systems may be avoided by considering events to be partially rather totally ordered. Occurrence nets, as described by Nielsen, Plotkin and Winskel have proved to be a natural structure for such an analysis, provided the infinite unfolding of a net can be truncated in a suitable manner. Here, we have extended the work of Nielsen *et al.* by providing such a method of truncation.

We have also observed that Petri nets can provide a model of asynchronous circuits that makes the partial ordering of events explicit. This made it possible to apply unfolding methods to hazard analysis. For a commonly studied example circuit (really a class of circuits of increasing size), the unfolding method reduced the cost of the analysis from exponential in the size parameter to polynomial. This demonstrates a possible advantage of unfoldings over state space search methods using partial orders, which in general are not effective for asynchronous circuits.

The problem of deadlock analysis has been shown to be hard, given the truncated unfolding, but solvable in practice using a branch and bound technique.

It is natural to speculate whether unfolding methods might be carried over to broader notions of verification, such as conformation checking [Dil88], temporal logic model checking [CE81b], or language containment of ω-automata [Kur86]. Conformation checking seems to be the most natural candidate, since it deals only with finite sequences. In such a program, trace structures might be modeled with Petri nets rather than finite automata, provided a suitable notion of parallel composition could be defined. Conformation testing could then be framed in terms of coverability. Building unfolding into the model checking or ω-automata frameworks is a more interesting problem, since these are concerned with properties of infinite runs, hence the notions of coverability and deadlock are not applicable. The expressive power of Petri nets will also undoubtedly prove to be a limitation. Extracting a partial order on events in more powerful models may well prove to be an interesting problem. Perhaps a more interesting problem would be to determine how partially ordered events relate to the symbolic model checking technique. Clearly, partially ordering of events implies some limit on mutual knowlegdge across a cut in a system, and this fact might be exploited in bounding the OBDD representation of its reachable states. In all events, there is much work left to be done.

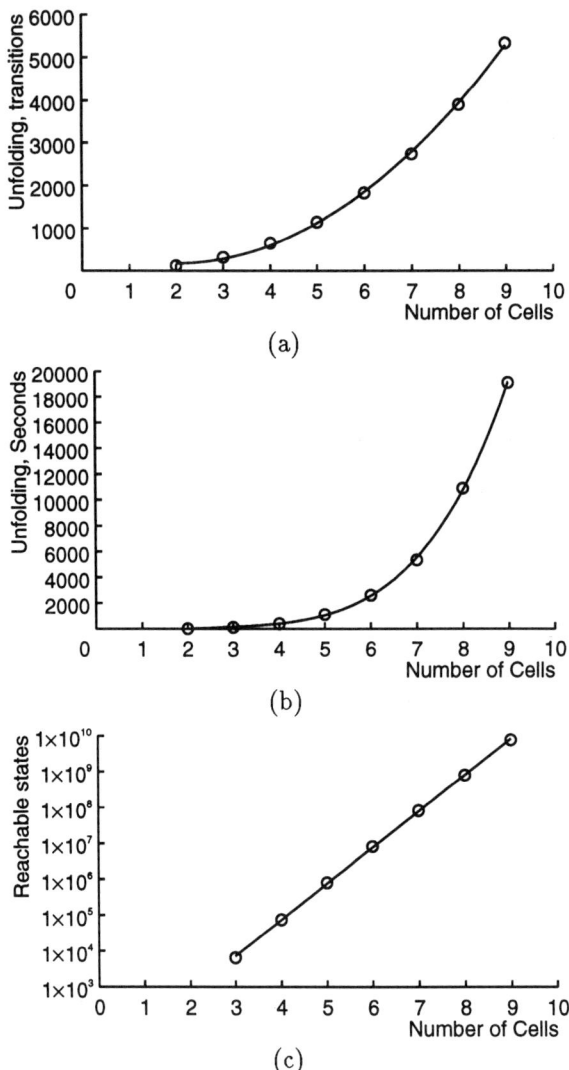

Figure 9.7 Performance of unfolding method on hazard-detection problem for the distributed mutual exclusion circuit

A partial order approach

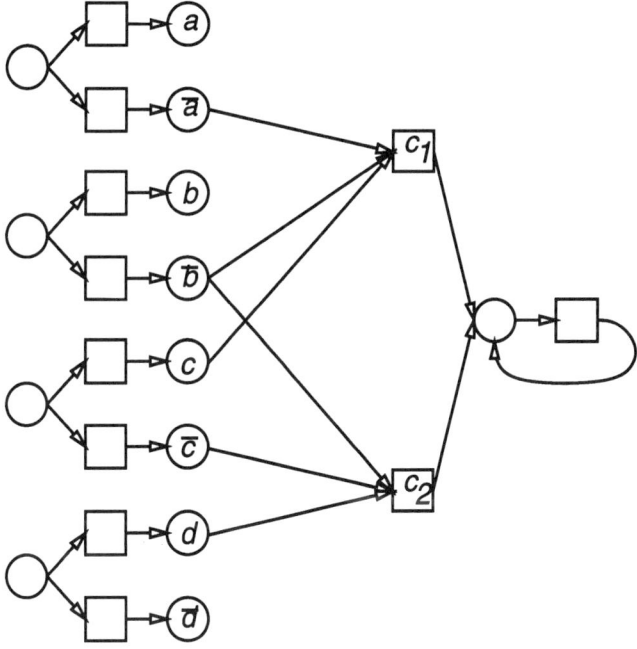

Figure 9.8 Reduction from 3-SAT problem to a terminal marking problem.

10
CONCLUSION

What we have seen in the preceding chapters is that Ordered Binary Decision Diagrams can be used as a representation in a wide variety of automatic verification algorithms, in order to cope with the state explosion problem. This can be done in a unified way by representing the algorithms in the Mu-Calculus fixed point notation. For fairly diverse families of regularly structured systems, the CTL model checking algorithm was observed to run in time and space which increased polynomially in the size of the system, while the number of reachable states increased exponentially. These results bear out a theoretical result bounding the OBDD representation of the transition relation for such systems. Standard automatic verification algorithms would be unsuitable for these examples because their complexity is proportional to the number of reachable states.

Using OBDD based techniques, and a language suitable for the abstract modeling of digital systems, it was possible to verify a fairly complex industrial design for a cache consistency protocol, finding a number of subtle errors in the process. The verification process is valuable not only because of the advantages of formalization and exhaustive checking, but because it can find protocol errors more quickly than simulation, despite the exponential growth in states as the model increases in size. The ability to isolate high level errors quickly shortens the loop between design and verification, making it possible to experiment more freely with alternative designs, and shortening the "critical path" from conceptualization to implementation.

By a technique of induction over processes, it is possible to prove properties of a protocol which are independent of the number of processes participating in the protocol. This type of proof requires a sufficient understanding of the protocol on the part of the designer to construct a process invariant. Invariants are difficult to find, but the symbolic model checker provides an aid in this process by producing counterexamples for unsound invariants. In the author's opinion, finding a process invariant for a protocol is not only of value as a proof technique – the understanding of the protocol required to formulate the invariant can lead to simpler and more elegant protocols. This is another reason for formalizing and verifying a protocol before attempting to implement it.

The verification technique based on occurrence nets shows that OBDDs are not the only representation that can be used to avoid the state explosion problem. There are, in fact, certain advantages to the occurrence net based method for the example presented, since the memory usage is small, and no heuristic technique is required to produce a variable ordering. Still, at this stage, the occurrence net method is certainly not as well advanced as the symbolic model checking method.

There are several areas where the current work falls short of the goal of complete automatic verification of digital systems. In the case of the Gigamax protocol, an abstract model of the protocol was verified and not the actual implementation. Verification of the implementation would have been impossible due to a lack of formal models of the components of the system (*ie.*, standard devices, such as memories, registers, programmable logic, central processing units, *etc.*). If such models were available from the manufacturers, in principle the methods described in chapter 7 could be used to show that the implementation is simulated by the abstract model. Hierarchical reasoning of this kind has been extensively studied by Kurshan [Kur87]. Unfortunately, simulation does not preserve existential CTL properties such as absence of deadlock. As mentioned previously, bisimulation equivalence, which preserves all CTL properties, is too strong for this purpose, since the abstract models are necessarily non-deterministic, and the actual implementation cannot (and should not) exhibit this non-determinism. A practical technique of abstraction which preserves existential CTL properties is needed if existential properties are to be proved using hierarchical reasoning.

There is also a need for heuristic strategies for generating process invariants in inductive proofs. Marelly and Grumberg view the design of the invariant as part of the design of the protocol. This is a useful point of view, but some automated help beyond the generation of counterexamples would be useful for this purpose.

Finally, this work concentrates on how to solve the verification problem, once it has been formalized as the satisfaction of a temporal logic formula by a finite

Conclusion 181

model, or as an appropriate relation between finite automata. There is, of course, a wide range of issues involved in formalizing the problem in the first place. For example, there is the ever present danger that the specification itself is incorrect. In the case of the very simple CTL formulas used to specify the Gigamax protocol, this is perhaps not a severe problem. The abstraction that was used to create a model for checking the sequential consistency property was, however, not obviously correct.

In general, there is a clear need for complete mechanical checking that the implementation of a processor or protocol matches the intended architecture (user model). This requires first of all a definitive model of the architecture – something that is currently lacking even for standardized architectures in the public domain. Second there must be a well defined criterion for determining what is a valid implementation of the architecture. Loosely, an implementation of a processor is equivalent to an architecture model if for all programs, the two machines produce the same "answer". However, for many reasons, this equivalence cannot be directly stated in terms of equivalence of finite state machines. For one, most modern CPU architectures have no explicitly defined notion of input and output. It is not adequate to view input and output as the sequence of loads or stores observed at the memory interface, since this sequence will differ among implementations (especially if the implementations contain cache memories, which is often the case). Solutions to the formalization problem are needed, but cannot be obtained by studying theoretical models alone. It is necessary to carefully consider what verification means in an engineering sense, as well as a mathematical sense.

Despite the shortcomings of current verification technology, it is clear that there are at least small areas of the problem space for which reasonable solutions exist, and these solutions can be put into practice to positive effect in an industrial setting. Those involved in verification research should perhaps take a closer look at engineering practice to determine how well the verification solutions match up with real engineering problems. This effort may lead not only to a more practical theory of formal verification, but also to a rich source of theoretical problems.

REFERENCES

[AB86] J. Archibald and J. L. Baer. Cache coherence protocols: Evaluation using a multiprocessor simulation model. *ACM Transactions on Computer Systems*, 4(4):273–298, 1986.

[Ake78] S. B. Akers. Binary decision diagrams. *IEEE Trans. Computers*, C-27(6):509–516, August 1978.

[BAMP81] M. Ben-Ari, Z. Manna, and A. Pnueli. The temporal logic of branching time. In *ACM Symp. Principles of Programming Languages*, pages 164–176, 1981.

[BBB+87] R. E. Bryant, D. Beatty, K. Brace, K. Cho, and T. Sheffler. COSMOS: A compiled simulator for MOS circuits. In *24th Design Automation Conference*, 1987.

[BBS90] Derek L. Beatty, Randal E. Bryant, and Carl-Johan H. Seger. Synchronous circuit verification by symbolic simulation: An illustration. In *Advanced Research in VLSI: Proceedings of the Sixth MIT Conference*, April 1990.

[BCD86] M. C. Browne, E. M. Clarke, and Dill. Automatic verification using temporal logic: Two new examples. In George J. Milne and P. A. Subramanyam, editors, *Formal Aspects of VLSI Design, Proceedings of the 1985 Edinburgh Workshop on VLSI*, pages 113–124. North-Holland, 1986.

[BCDM86] M. C. Browne, E. M. Clarke, D. L. Dill, and B. Mishra. Automatic verification of sequential circuits using temporal logic. *IEEE Transactions on Computers*, C-35(12):1035–1044, 1986.

[BCG86] M.C. Browne, E. M. Clarke, and O. Grumberg. Reasoning about networks with many identical finite state processes. In *ACM Symp. Principles of Distributed Computing 5*, 1986.

[BCG87] M.C. Browne, E. M. Clarke, and O. Grumberg. Characterizing kripke structures in temporal logic. Technical Report 87-104, Carnegie-Mellon University. Computer Science Department, 1987.

[BCL] J. R. Burch, E. M. Clarke, and D. E. Long. Symbolic model checking with partitioned transition relations. To appear in the Proceedings of VLSI'91.

[BCL91a] Jerry R. Burch, Edmund M. Clarke, and David E. Long. Representing circuits more efficiently in symbolic model checking. In *Proceedings of the 28th ACM/IEEE Design Automation Conference*, San Francisco, California, June 1991.

[BCL91b] Jerry R. Burch, Edmund M. Clarke, and David E. Long. Symbolic model checking with partitioned transition relations. In A. Halaas and P. B. Denyer, editors, *Proceedings of the IFIP International Conference on Very Large Scale Integration*, Edinburgh, Scotland, August 1991.

[BCM+90] J. R. Burch, E. M. Clarke, K. L. McMillan, D. L. Dill, and J. Hwang. Symbolic model checking: 10^{20} states and beyond. In *Proceedings of the Fifth Annual Symposium on Logic in Computer Science*, June 1990.

[BF89a] S. Bose and A. Fisher. Verifying pipelined hardware using symbolic logic simulation. In *IEEE International Conference on Computer Design*, 1989.

[BF89b] Soumitra Bose and Allan L. Fisher. Automatic verification of synchronous circuits using symbolic logic simulation and temporal logic. In Luc Claesen, editor, *Proceedings of the IMEC-IFIP International Workshop on Applied Formal Methods For Correct VLSI Design*, pages 759–764, November 1989.

[Bil87] J. P. Billon. Perfect normal forms for discrete functions. Technical Report 87019, BULL, March 1987.

[Bry86] R. E. Bryant. Graph-based algorithms for boolean function manipulation. *IEEE Transactions on Computers*, C-35(8), 1986.

[Bry88] Randal E. Bryant. Verifying a static ram design by logic simulation. In Jonathan Allen and F. Thomson Leighton, editors, *Advanced Research in VLSI: Proceedings of the Fifth MIT Conference*, pages 335–349. MIT Press, 1988.

[Bry91] R. E. Bryant. On the complexity of VLSI implementations and graph representations of boolean functions with application to integer multiplication. *IEEE Trans. Computers*, 40(2):205 – 213, February 1991.

[BS90] Randal E. Bryant and Carl-Johan Seger. Formal verification of digital circuits using symbolic ternary system models. In Robert Kurshan and

REFERENCES 185

Edmund M. Clarke, editors, *Workshop on Computer-Aided Verification*, New Brunswick, New Jersey, June 1990. Center for Discrete Mathematics and Theoretical Computer Science (DIMACS).

[Bur84] J. P. Burgess. Basic tense logoc. In D. Gabbay and F. Guenthner, editors, *Handbook of Philosophical Logic. Volume II: Extensions of Classical Logic*, pages 89–134. D. Reidel, 1984.

[CBM89] Olivier Coudert, Christian Berthet, and Jean Christophe Madre. Verification of synchronous sequential machines based on symbolic execution. In Joseph Sifakis, editor, *Automatic Verification Methods for Finite State Systems, International Workshop, Grenoble, France*, volume 407 of *Lecture Notes in Computer Science*. Springer-Verlag, June 1989.

[CDK90] E. M. Clarke, I. A. Draghicescu, and R. P. Kurshan. A unified approach for showing language containment and equivalence between various types of ω-automata. In *Fifteenth Colloquium on Trees in Algebra and Programming*, volume 431 of *Lecture Notes in Computer Science*, Copenhagen, Denmark, May 1990. Springer-Verlag.

[CE81a] E. M. Clarke and E. A. Emerson. Characterizing properties of parallel programs as fixpoints. In *Seventh International Colloqium on Automata, Languages, and Programming*, volume 85 of *LNCS*, 1981.

[CE81b] E. M. Clarke and E. A. Emerson. Synthesis of synchronization skeletons for branching time temporal logic. In Dexter Kozen, editor, *Logic of Programs: Workshop*, volume 131 of *Lecture Notes in Computer Science*, Yorktown Heights, New York, May 1981. Springer-Verlag.

[CES86] E. M. Clarke, E. A. Emerson, and A. P. Sistla. Automatic verification of finite-state concurrent systems using temporal logic specifications. *ACM Transactions on Programming Languages and Systems*, 8(2):244–263, 1986.

[CGK89] E. M. Clarke, O. Grumberg, and R. P. Kurshan. A synthesis of two approaches for verifying finite state concurrent systems. In *Logic at Botik '89, Symposium on Logical Foundations of Computer Science*, volume 363 of *Lecture Notes in Computer Science*. Springer-Verlag, July 1989.

[CGL92] Edmund M. Clarke, Orna Grumberg, and David E. Long. Model checking and abstraction. In *Proceedings of the Annual ACM Symposium on Principles of Programming Languages*, January 1992.

[CHPP87] P. Caspi, N. Halbwachs, D. Pilaud, and J. A. Plaice. Lustre, a declarative language for programming synchronous systems. In *14th ACM Symp. on Principles of Programming Languages*, January 1987.

[CLM89a] E. M. Clarke, D. E. Long, and K. L. McMillan. Compositional model checking. In *Proceedings of the Fourth Annual IEEE Symposium on Logic in Computer Science*, 1989.

[CLM89b] E. M. Clarke, D. E. Long, and K. L. McMillan. A language for compositional specification and verification of finite state hardware controllers. In *9th International Symposium on Hardware Description Languages and their Applications*, 1989.

[CM88] K. M. Chandy and J. Misra. *Parallel program design : a foundation*. Addison-Wesley, 1988.

[CMB91] Olivier Coudert, Jean Christophe Madre, and Christian Berthet. Verifying temporal properties of sequential machines without building their state graphs. In E. M. Clarke and R. P. Kurshan, editors, *Computer Aided Verification, '90*, volume 3 of *DIMACS Series in Discrete Mathematics and Theoretical Computer Science*, 1991.

[DC86] D. L. Dill and E. M. Clarke. Automatic verification of asynchronous circuits using temporal logic. *IEE Proceedings, Pt. E*, 133(5):276–282, September 1986.

[Dil88] D. Dill. Trace theory for automatic hierarchical verification of speed-independent circuits. Technical Report 88-119, Carnegie Mellon University, Computer Science Dept, 1988.

[EL86] E. A. Emerson and C.-L. Lei. Efficient model checking in fragments of the propositional mu-calculus. In *Proceedings of the Second Annual Symposium on Logic in Computer Science*. IEEE Computer Society Press, June 1986.

[FB89] Allan L. Fisher and Randal E. Bryant. Performance of COSMOS on the IFIP workshop benchmarks. In *Proceedings of IMEC Conference*, 1989.

[FMK90] M. Fujita, Y. Matsunaga, and T Kakuda. Automatic and semi-automatic verification of switch-level circuits with temporal logic and binary decision diagrams. In *ICCAD*, pages 38–41, 1990.

[GL91] Orna Grumberg and David E. Long. Model checking and modular verification. In J. C. M. Baeten and J. F. Groote, editors, *Proceedings of CONCUR '91: 2nd International Conference on Concurrency Theory*, volume 527 of *Lecture Notes in Computer Science*. Springer-Verlag, August 1991.

[God90] P. Godefroid. Using partial orders to improve automatic verification methods. In *Workshop on Computer Aided Verification*, 1990.

[GS] S. M. German and A. P. Sistla. Reasoning about systems with many processes. GTE Laboratories Inc., Waltham, Massachusetts.

[GS91] Susanne Graf and Bernhard Steffen. Compositional minimization of finite state systems. In Robert Kurshan and Edmund M. Clarke, editors, *Computer-Aided Verification, Proceedings of the 1990 Workshop*, volume 3 of *DIMACS Series in Discrete Mathematics and Theoretical Computer Science*. American Mathematical Society, 1991.

[GW91] P. Godefroid and P. Wolper. A partial approach to model checking. In *LICS*, 1991.

[Hoa69] C. A. R. Hoare. An axiomatic basis for computer programming. *Comm. ACM*, 12:576–580,583, October 1969.

[KC90] S. Kimura and E. M. Clarke. A parallel algorithm for constructing binary decision diagrams. In *1990 IEEE International Conference on Computer Design*, September 1990.

[KM89] R. Kurshan and K. L. McMillan. A structural induction theorem for processes. In *ACM Symposium on Principles of Distributed Computing*, Edmonton, Alberta, 1989.

[Kro87] Fred Kroger. *Temporal Logic of Programs*. Springer-Verlag, 1987.

[Kur85] R. P. Kurshan. Modelling concurrent processes. In *Symp. in Applied Math. 31*, pages 45–57. 1985.

[Kur86] R. P. Kurshan. Testing containment of ω-regular languages. Technical Report 1121-861010-33-TM, Bell Laboratories, 1986.

[Kur87] R. P. Kurshan. Reducibility in analysis of coordination. In *LNCS*, volume 103, pages 19–39. Springer-Verlag, 1987.

[LN91] B. Lin and A. R. Newton. Efficient symbolic manipulation of equivalence relations and classes. In *IMEC-IFIP International Workshop on Formal Methods in VLSI Design*, Miami, Florida, 1991.

[LP85] Orna Lichtenstein and Amir Pnueli. Checking that finite state concurrent programs satisfy their linear specification. In *Conference Record of the Twelfth Annual ACM Symposium on Principles on Programming Languages*, 1985.

[LTN90] B. Lin, H. J. Touati, and A. R. Newton. Don't care minimization of multi-level sequential logic networks. In *ICCAD*, pages 414–417, 1990.

[Mar85] A. J. Martin. The design of a self-timed circuit for distributed mutual exclusion. In Henry Fuchs, editor, *1985 Chapel Hill Conference on VLSI*, pages 245–260. Computer Science Press, 1985.

[MB59] D. E. Muller and W. S. Bartky. A theory of asynchronous circuits. In *The Annals of the Computation Laboratory of Harvard University. Volume XXIX: Proceedings of an International Symposium on the Theory of Switching, Part I*, pages 204–243. Harvard University Press, 1959.

[Mil80] R. Milner. *A Calculus of Communicating Systems*, volume 92 of *Lecture Notes in Computer Science*. Springer-Verlag, 1980.

[Mil83] G. J. Milne. Circal, calculus for circuit descriptions. *Integration*, 1:121–160, 1983.

[MO81] Y. Malachi and S. S. Owicki. Temporal specifications of self-timed systems. In H. T. Kung, B. Sproull, and G. Steele, editors, *VLSI Systems and Computations*. 1981.

[MP81] Z. Manna and A. Pnueli. Verification of concurrent programs: the temporal framework. In R. S. Boyer and J. S. Moore, editors, *The Correctness Problem in Computer Science*, pages 215–273. 1981.

[MS91] K.L. McMillan and J. Schwalbe. Formal verification of the Encore Gigamax cache consistency protocol. In *International Symposium on Shared Memory Multiprocessors*, 1991.

[NH84] R. De Nicola and M. Hennessy. Testing equivalences for processes. *Theoretical Computer Science*, 34(83), 1984.

[NPW81] M. Nielsen, G. Plotkin, and G. Winskel. Petri nets, event structures and domains, part I. *Theoretical Computer Science*, 13:85–108, 1981.

[Par74] David Park. Finiteness is mu-ineffable. Theory of Computation Report No. 3, The University of Warwick, 1974.

[PL89] D. K. Probst and H. F. Li. Abstract specification, composition, and proof of correctness of delay-insensitive circuits and systems. Technical report, Concordia University, Dept. of Computer Science, 1989.

[PL90] D. K. Probst and H. F. Li. Using partial order semantics to avoid the state explosion problem in asynchronous systems. In *Workshop on Computer Aided Verification*, 1990.

[PL91] D. K. Probst and H. F. Li. Partial order model checking: A guide for the perplexed. In *Third Workshop on Computer Aided Verification*, pages 405–416, July 1991.

[Pnu77] A. Pnueli. The temporal semantics of concurrent programs. In *18th Symposium on Foundations of Computer Science*, 1977.

[Pnu86] A. Pnueli. Applications of temporal logic to the specification and verification of reactive systems: A survey of current trends. In *Lecture Notes in Computer Science*, volume 224, pages 510–584. Springer-Verlag, 1986.

[QS81] J. P. Quielle and J. Sifakis. Specification and verification of concurrent systems in CESAR. In *Proceedings of the Fifth International Symposium in Programming*, 1981.

[RU71] N. Rescher and A. Urquhart. *Temporal Logic*. Springer-Verlag, 1971.

[Sei80a] C. L. Seitz. Ideas about arbiters. *Lambda*, 10(14), 1980.

[Sei80b] C. L. Seitz. System timing. In Carver Mead and Lynn Conway, editors, *Introduction to VLSI Systems*, pages 218–262. Addison-Wesley, 1980.

[SG89] Z. Shtadler and O. Grumberg. Network grammars, communication behaviors and automatic verification. In *Workshop on Automatic Verification Methods for Finite State Systems*, LNCS, pages 151–165. Springer-Verlag, 1989.

[Smu68] R. M. Smullyan. *First Order Logic*. Springer-Verlag, 1968.

[Tar55] A. Tarski. A lattice-theoretical fixpoint theorem and its applications. *Pacific J. Math.*, 5:285–309, 1955.

[TBK91] H. J. Touati, R. K. Brayton, and R. P. Kurshan. Testing language containment for ω-automata using BDD's. In *IMEC-IFIP International Workshop on Formal Methods in VLSI Design*, Miami, Florida, 1991.

[Tho84] R. H. Thomason. Combinations of tense and modality. In D. Gabbay and F. Guenthner, editors, *Handbook of Philosophical Logic. Volume II: Extensions of Classical Logic*, pages 89–134. D. Reidel, 1984.

[TSL+90] H. J. Touati, H. Savoj, B. Lin, R. K. Brayton, and A. Sangiovanni-Vincentelli. Implicit state enumeration of finite state machines using BDD's. In *ICCAD*, pages 130–133, 1990.

[Val89] A. Valmari. Stubborn sets for reduced state space generation. In *10th Int. Conf. on Application and Theory of Petri Nets*, 1989.

[Val90] A. Valmari. A stubborn attack on the state explosion problem. In *Workshop on Computer Aided Verification*, 1990.

[vdS83] Jan L. A. van de Snepscheut. *Trace Theory and VLSI design*. PhD thesis, Department of Computer Science, Eindhoven University of Technology, October 1983.

[WL89] P. Wolper and V Lovinfosse. Verifying properties of large sets of processes with network invariants. In *Workshop on Automatic Verification Methods for Finite State Systems*, LNCS, pages 68–80. Springer-Verlag, 1989.

[Wol83] Pierre Wolper. Temporal logic can be more expressive. *Information and Control*, 56:72–99, 1983.

[YTK91] Tomohiro Yoneda, Yoshihiro Tohma, and Yutaka Kondo. Acceleration of timing verification method based on time Petri nets. *Systems and Computers in Japan*, 22(12):37–52, 1991.

INDEX

Abstract model, 89
Abstraction, 108, 127
Actual parameters, 75, 84
AndExists, 37
Apply, 36
Arbiter, 39, 49, 54
ASSIGN keyword, 63, 72, 82
Assignment declaration, 72
Asymptotic performance, 47, 106, 109, 123
Asynchronous circuits, 44, 66, 168
Atomic proposition, 17
Automata theoretic methods, 5
Axioms, 13
Backward closed, 160
Base functions, 81
Binary Decision Diagrams, 3, 32
Bisimulation, 6, 119
Boolean algebra, 18, 25, 33
Boolean formulas, 114
Boolean quantification, 27
Boolean representations, 25
Bottom-up substitution, 146
Bounded knowledge, 106
Bounded tree-width, 56
Bounded width, 49
Branching time, 15
Bus arbiter, 39
Bus snooping, 87, 89
Büchi automata, 120
Cache blocks, 87
Cache consistency protocol, 49, 87
Cache replacement policy, 91
Cache, 89
Call-by-name, 76

Call-by-reference, 76, 84
Canonical form, 32
Case expression, 63, 71
CCS, 6
  and induction, 139
Circuit algebra, 6
Cluster, 87
Compiler, 78
Complexity
  of AndExists, 38
  of symbolic model checking, 38
Compose, 37, 116, 122, 144
Compositional model checking, 8
Computation Tree Logic, 16
Computation tree, 17
Configuration, 160
Conformance, 6
Conjunctive partitioning, 123
Constrain, 124, 126
Continuous, 19
Copy back, 89
Counterexample, 101, 105, 135
Courtesy read, 98
Coverable, 167
Cross section, 52
CTL, 16, 61, 73
Cutoff point, 164
Data dependencies, 65
Data structure, 65
Dataflow, 62
Deadlock, 62, 100–101
  and occurrence nets, 171
DEFINE keyword, 65, 75, 83
Denotation, 79
Dining philosophers, 165

Directed acyclic graphs, 33
Discrete time, 14
Distributed mutual exclusion, 39, 45
DME, 45, 49, 54, 123, 168
Domain partitioning, 124, 147
Dynamic hazard, 168
Early quantification, 45, 122, 126
Enumerated types, 61–62, 71, 79
Equivalence relation, 126, 143
Equivalence, 123, 143
Event structure, 155
Expressions, 69, 80
FAIR keyword, 74, 84
Fairness constraint, 67, 74, 79, 85, 118
Fairness, 62
Final state, 162
Finite model, 17
Finite state machines, 62
  asynchronous, 44
  synchronous, 39
Firing sequences, 159
Fixed points, 17–19
Floyd-Hoare logic, 12
Formal parameters, 75, 84
Formal verification, 1
Formally monotonic, 114
Forward cross section, 52
Forward width, 52
Frame, 13
Functional, 18
Gigamax, 87, 134
Global bus, 87
GORMEL, 141
Hidden weighted bit, 54
Hit, 89
Homomorphic reductions, 7
Identifier, 70, 76
Image, 117
Inclusion operator, 71
Indexed CTL, 140
Individual interpretation, 114

Individual variables, 114
Induction, 8, 129
INIT keyword, 68, 73, 84
Initial condition, 79
Initial state, 63
Instance, 65, 72, 75
Instantiation, 83
Interleaving model, 44, 46, 49, 123
Internal node, 33
Invalid state, 90
Invalidate, 89, 91
Invariant, 129
Iterative abstraction, 146
Iterative squaring, 39
Kripke model, 21, 27, 63
Language containment, 6, 120, 126
  and induction, 139
Linear temporal logic, 14
Livelock, 101
Liveness properties, 5
Liveness, 62
Local configuration, 163
LUSTRE, 62
Marking, 158
Mealy machine, 61, 69, 143
Metastability, 12
Miss, 89
Modal logics, 13
Modal tense logic, 16
Model checking problem, 18
Model checking, 2, 4, 17
Model theory, 13, 16
Model, 13
Modified breadth first search, 123
MODULE keyword, 75, 83
Module, 72, 75
Monotonic, 19
Mu-Calculus, 7, 113
Network grammars, 140
Next keyword, 70
Non-determinism, 15, 88
Non-deterministic choice, 61, 64

Index

OBDD, 31, 42
Observational equivalence, 6
Occurrence net, 155, 159
Ordered Binary Decision Diagram, 31
Ordered decision tree, 31
Owned state, 90
Parallel assignment, 62
Partially ordered model, 108
Partitioned transition relations, 122
Path quantifiers, 62
Path, 18
Persistence, 154
Petri net, 155, 158
Pomtree, 155
Postset, 158, 162
Pre-order, 130
Precedence, 70, 74
Prefix operator, 132
Preset, 158
Process closure, 140
Process invariant, 130
Process keyword, 72, 77, 84
Process quantifiers, 8, 140
Process, 66, 77, 84
PSPACE, 39
Queues, 95
Range partitioning, 124
Reachable states, 42, 48, 117, 125
Reactive, 3, 12, 17
Read, 91
Read-owned, 91
Read-shared, 91
Reduce, 32
Reduction, 7
Relational formulas, 114
Relational interpretation, 114
Relational variables, 114
Replacement, 89
Reply-owned, 92
Reply-stall, 92
Reply-waiting, 92

Response, 91
Restrict, 48, 125, 145
Reverse cross section, 52
Reverse width, 52
Running keyword, 67, 77
Safety, 62, 100
Semantics, 13
  of CTL, 18
  of SMV, 78
  syntax-directed, 61, 79
Sequential consistency, 100
Sequential substitution, 148
Set expression, 71
Shared memory multiprocessor, 87
Shared state, 90
Simulation, 87, 120, 132
Simultaneous model, 44, 46, 67
Since, 14
SMV, 61
Snoop, 92
SPEC keyword, 64, 73, 84
Specification, 63, 73, 79, 100
Speed-independent, 4, 45
Split transaction bus, 89
State explosion problem, 2
State minimization, 8
Street automata, 120
Stuttering, 141
Substitution, 146
Symbolic model checker, 87
Symbolic model checking, 3, 36, 42, 61, 113, 122
Symbolic models, 26
Symbolic simulation, 126
Symmetric functions, 47
Syntax
  of CTL, 17
  of SMV, 62
  of the Mu-Calculus, 114
Syntax-directed, 79
Tableau, 5, 7
Temporal logic, 2, 13